Albert Einstein
Memorial Lectures

Albert Einstein
Memorial Lectures

Editors

Jacob D. Bekenstein
Raphael Mechoulam

The Israel Academy
of Sciences and Humanities

World Scientific

Published by

The Israel Academy of Sciences and Humanities
Albert Einstein Square
P.O. Box 4040
Jerusalem 91040
Israel

and

World Scientific Publishing Co. Pte. Ltd.

5 Toh Tuck Link, Singapore 596224

USA office: 27 Warren Street, Suite 401-402, Hackensack, NJ 07601

UK office: 57 Shelton Street, Covent Garden, London WC2H 9HE

ALBERT EINSTEIN MEMORIAL LECTURES

ISBN-13 978-981-4329-42-2
ISBN-10 981-4329-42-8
ISBN-13 978-981-4329-43-9 (pbk)
ISBN-10 981-4329-43-6 (pbk)

Printed in Singapore by Mainland Press Pte Ltd.

Preface

Albert Einstein was a leader in establishing the foundations of scientific research in Israel. In the 1920s, when the Hebrew University of Jerusalem — the first university in what was then British-Mandate Palestine — was founded, Einstein was elected Head of the Academic Council of the Board of Governors. Both the administration and the staff assumed that he would be a distinguished figurehead, but, to everyone's surprise, he became very involved in the new university's activity. While its American financial backers presumably had envisioned something on the model of an American-type college, he aspired for the Hebrew University to follow in the tradition of the great European research universities. Einstein's insistence led to the formation of an international academic committee, which duly visited Jerusalem and strongly supported his understanding of how the university should develop. This conception became the intellectual basis not only of the Hebrew University, but of all the universities established thereafter in Israel. The Technion in Haifa, founded before the Hebrew University as a technical college, was also gradually transformed into a research institution.

Einstein's relations with the Hebrew University were turbulent at times, but he maintained them until the end of his life, and he bequeathed his archives to the university. Several of his students and colleagues joined the faculties of Israeli academic institutions.

Einstein was not only one of the outstanding scientists of the twentieth century; he was also a symbol of intellectual independence and leadership. Unlike most German academics, he refused to support the German entry into World War I. He was one of the first academics to recognize the danger of Nazism, and most major scientists followed his call. Notwithstanding his deeply rooted pacifism, he feared that the West would be preempted by Nazi Germany in developing atomic energy for military purposes, leading him to initiate its development in the United States by penning his famous letter to President Roosevelt.

David Ben-Gurion invited Einstein to become the first President of the State of Israel — an offer he turned down. He valued his independence and could not envision himself becoming absorbed in official receptions and duties of state.

For us, as scientists, he represents the ultimate personal ideal — a researcher who opened new vistas in science and an intellectual who led by personal example, without the need of official titles or positions.

With the annual Albert Einstein Memorial Lectures, the Israel Academy of Sciences and Humanities conveys its aspiration to sustain the legacy of this intellectual giant.

The editors thank, first and foremost, the contributors, for allowing us to publish their lectures. We are also grateful to Professor Yohanan Friedmann and the Publications Committee of the Israel Academy of Sciences and Humanities, which he headed until recently, for their guidance of this project; to the Director of the Academy's Publications Department, Ms. Tali Amir; and especially to Ms. Deborah Greniman, its Senior Editor of English-language Publications, for her inestimable help. Ms. Esther Rosenfeld gave the volume an expert round of proofreading. Our thanks, finally, to World Scientific Publishing for proposing the publication of this volume and for their care and cooperation in producing it.

Jacob Bekenstein and Raphael Mechoulam
Jerusalem, July 2011

List of Contributors

Timothy Gowers wrote his Ph.D. on the geometry of Banach spaces and went on to solve several old problems in that area, some due to Banach himself. In 1998 he received a Fields Medal, partly for this work and partly for a highly influential new proof of Szemerédi's theorem. He is currently a Royal Society Research Professor at Cambridge University.

Haim Harari is Annenberg Professor of High-Energy Physics at the Weizmann Institute and Chairman of the Davidson Institute of Science Education. He is a member of the Israel Academy of Sciences and the American Academy of Arts and Sciences, and recipient of, among others, the Rothschild Prize in Physics (Yad Hanadiv, 1976), the Israel Prize, the EMET Prize in Education (A.M.N. Foundation for Science, Art and Culture, sponsored by the Prime Minister of Israel, 2004), and the Harnack Medal (Max Planck Society, 2001). He was President of the Weizmann Institute (1988–2001), Chairman of the Planning and Budgeting Committee of Israel's Council for Higher Education (1979–1985), and co-founder of Perach, a national tutoring program for underprivileged children in Israel.

Roger Kornberg is Winzer Professor of Medicine in the Department of Structural Biology at Stanford University. As a postdoctoral fellow and member of the scientific staff at the Laboratory of Molecular Biology in Cambridge, England (1972–1975), he discovered the nucleosome, the basic unit of DNA coiling in chromosomes. He is the recipient of the Welch Award in chemistry (2001), the Leopold Mayer Prize in biomedical sciences (French Academy of Sciences, 2002) and the Nobel Prize in Chemistry (2006).

Yuan T. Lee, former President of the Academia Sinica (1994–2006), was co-recipient of the Nobel Prize in Chemistry (1986) for his contributions concerning the dynamics of chemical elementary processes, particularly his development of the method of crossed molecular beams for the study of important reactions for relatively large molecules. He is a member of the (U.S.) National Academy of Sciences and of Academia Sinica, and the recipient of, among others, the National Medal of Science (awarded by the President of the United States, 1986), the Peter Debye Award (American Chemical Society, 1986) and the Faraday Medal (Royal Society of Chemistry, London, 1992).

Jean-Marie Lehn is Director of the Institut de Science et d'Ingénierie Supramoléculaires (ISIS) in Strasbourg and Professor of Chemistry at the Collège de France. He was awarded the Nobel Prize in Chemistry (1987), together with Charles J. Pedersen and Donald J. Cram, for their development and use of molecules with structure-specific interactions of high selectivity.

Carlo Rubbia is best known for his discovery of W and Z particles, for which he was awarded the Nobel Prize in Physics (1984), and for his proposal for a safe nuclear reactor in the form of a coupled accelerator-reactor using thorium as fuel. He has been instrumental in the development and operation of the European Laboratory for Particle Physics (CERN) in Geneva and served as the Laboratory's Director-General for five years (1989–1993).

Shlomo Sternberg is George Putnam Professor of Pure and Applied Mathematics at Harvard University. Best known for his work on symplectic geometry, he is a Fellow of the American Academy of Arts and Sciences, the (U.S.) National Academy of Sciences and the Royal Academy of Spain, and he has been a Visiting Fellow of the Mortimer and Raymond Sackler Institute of Advanced Studies, Tel Aviv University.

John E. Wansbrough (1928–2002) was Professor of Semitic Studies and Pro-Director at the School of Oriental and African Studies of London University. He revolutionized the scholarly study of Islamic origins with his groundbreaking, controversial works *Quranic Studies* (1977) and *The Sectarian Milieu* (1978), in which he argued that the Koran, like the Jewish and the Christian Bible, was the product of a long period of gestation, resulting in a distinctive version of the biblical tradition of monotheism, on the same historical and theological footing as its predecessors.

Steven Weinberg is Professor of Physics and Astronomy at the University of Texas, where he holds the Jack S. Josey–Welch Foundation Regents Chair in Science, and was formerly Higgins Professor of Physics at Harvard (1973–1982). His honors include the Nobel Prize in Physics (1979) and the National Medal of Science (awarded by the President of the United States, 1991), election to numerous academies and sixteen honorary doctoral degrees.

Contents

What Can Pure Mathematics Offer to Society?

W. Timothy Gowers

It is a great pleasure and honor for me to be here this evening. The word honor speaks for itself, but perhaps not everyone here knows why it is such a pleasure. It is because, for one reason or another, my own mathematical life has had a lot to do with Israel. It is not often that one goes to a conference that changes the course of one's mathematical life. But I have been to one such conference, and that was here in Jerusalem. Furthermore, many Israeli mathematicians have been very kind to me in a number of different ways, so it is extremely nice for me to be back here on this occasion.

Let me start with a little discussion of what pure mathematics actually is before going on to what it has to offer to society. Let's look at three traditional attitudes to the question of what mathematics is. One is the Platonist attitude. It holds that mathematics is a discourse about some sort of objective reality, albeit a somewhat abstract reality. Platonists would believe that if I say $2 + 2 = 4$, what I mean is that the number 2, which has some objective existence, when added to itself, really does give you 4, which has some other objective existence. So if you ask a question in mathematics, then the answer would be yes or no, and, objectively, the statement would be true or false.

A second attitude is the logicist one. In mathematics, this was a program that attempted to justify our confidence in mathematical statements, by showing that they could be deduced, using just the rules of logic, from just a few axioms, assumptions whose truth was supposed to be so obvious as to be completely undeniable. So if, by using logical rules that are obviously valid, you can get from things that everybody would have to be mad to deny to some complicated statement, then that gives you some confidence that the complicated statement really is true.

A third possible attitude is the formalist position, which, roughly speaking, is that mathematics is sort of a game that you play with marks on a piece of paper. If you have got some marks written down, and you have

1

precisely defined rules that allow you to replace one set of marks with another, and if the first set of marks represents some statements that everybody accepts, then the second set of marks somehow represents what you are hoping to demonstrate. To put it in a flippant way, if you apply the rules correctly, then you can get your paper published. It doesn't really matter whether the marks mean anything; all that matters is that you apply the rules correctly.

Now, let me try to say a little about the advantages and disadvantages of these various points of view. Why should people like Platonism? One reason is that if you look at the mathematical statement $2 + 2 = 4$, it sounds as if it is saying something about something. It resembles the statement 'that chair is next to this chair.' When I am talking about chairs, I am confident that I am talking about something. For similar reasons, when I talk about a mathematical statement like $2 + 2 = 4$, it feels as though I am talking about something. One difference is that I can't 'see' the number 2, but we can explain that by saying that 2 is something abstract. You can't actually feel it directly, but you somehow have access to it in another way. All this affects the way one actually thinks about mathematics. If you try to visualize a number, you have a feeling that you are visualizing something, that there is something to visualize.

However, this abstract world really is quite hard to pin down, as can be seen most clearly in situations in which two mathematicians have philosophical disagreements. For example, suppose you ask whether the number infinity actually *exists* or not? Mathematician A thinks yes, infinity exists, but mathematician B thinks it doesn't exist. There seems to be no way of resolving the dispute. You can't take a trip into the abstract realm and try to find the number infinity; it just doesn't work like that. There are some more serious objections of a similar kind to Platonism, but they apply to logicism as well, so let me move on to that.

Part of the appeal of logicism is that the logicist program was much more successful than anyone had the right to expect. In the nineteenth century, mathematicians struggled to put various branches of mathematics on a firm footing. For example, at the early stages there were serious problems with getting calculus to look logically valid, as several of its assumptions seemed rather dubious. Eventually this was corrected by a process which, although it has its roots as far back as Euclid's axiomatic approach to geometry and other subjects, really took off only at the end of the nineteenth and the beginning of the twentieth century, in particular with the work of Bertrand Russell and Alfred Whitehead. They showed that all of conventional mathematics can be reduced to statements about sets, and in particular to some axioms about sets whose truth seems, in most cases, to be very hard to deny. So in a sense the whole of mathematics reduces to some completely obvious statements about sets.

So why should one have any doubts about that at all? Well, partly because the actual choice of axioms that you start with has a lot of leeway; you can choose them in all sorts of different ways. Some of the axioms themselves are not quite as obviously true as others, and some seem to have been constructed rather artificially in order to make the whole program work. Nevertheless, according to a logicist, you can in a sense reduce statements about numbers to statements about sets (if you are prepared to do some funny things to sets), though in the process the former statements become much harder to understand. It is not clear that this process is really doing justice to what we do when we think about numbers, but the logicists would claim that it is not their goal to do justice to our psychology.

There is also a much more serious problem with logicism than the artificiality of any system of axioms for set theory. Kurt Gödel showed that one cannot come up with a single system of axioms to do everything that is required. This connects with a famous unsolved problem in mathematics: Is the continuum hypothesis true or not? Georg Cantor famously proved that the (infinite) number of real numbers is larger than the (also infinite) number of whole numbers. But this left open the question of whether you can find some set that is bigger than the set of whole numbers but smaller than the set of real numbers — an intermediate infinity, so to speak. This is a very natural question; the statement that no intermediate level of infinity exists is called the continuum hypothesis. Gödel showed that you can never disprove this statement: It is consistent with the other axioms of set theory that there be no intermediate level of infinity. Then, in the 1960s, Paul Cohen showed that the negation of the continuum hypothesis is also consistent with the axioms of set theory. So you can neither prove nor disprove the continuum hypothesis.

This is surprising. It is a serious difficulty, I think, both for Platonism and, in a sense, for logicism — if you want to use logicism as a means to being absolutely confident in the truth of any given mathematical statement. I think most professional mathematicians, if asked, 'is the continuum hypothesis true?' would respond: 'Well, either you can accept it and then you can prove this, that and the other, or you can deny it and then you can prove different things, and that's all you can say.'

So what about formalism? Well, the obvious advantage of formalism is that it avoids the difficulty of whether the statements are true or false. The formalist would take exactly the attitude I have just described: It doesn't really matter whether the statement is true or false; all we care about is what we can deduce from whatever set of assumptions we pick, using the rules we allow ourselves. Typically, we start with a statement, represented as a string of symbols, and apply to it replacement rules that tell us which strings of symbols may be replaced with which other ones.

What could possibly be interesting about this strange game? One answer is that often when you play it, you find that in order to get from your starting assumptions to your conclusion, you have to pass through an intermediate stage where there is a very long string of symbols. You start with something short, and you are trying to reach something else that is short, but you get enormously long strings of symbols in between. In fact, not only is it often very difficult to find a route from start to finish, but in a certain precise sense it is impossible to do so systematically. You can't tell a computer: 'Here is a theorem I want to prove; here are the assumptions; here are the rules of logic — now just grind away and prove it for me.' Whatever algorithm you put into the computer, somebody can devise some clever mathematical statement with a proof that the computer will never find.

This very fact makes mathematics an enjoyable thing to do. It's difficult to get from your assumptions to your conclusions, and you don't always know whether you are going to succeed. That brings an excitement to the whole business of doing mathematics, even on a very abstract level. Obviously, doing mathematics has other rewards than just watching your intermediate complicated results suddenly collapse into something simple, but that is one of the things that is rather nice about it.

If you speak to any mathematician, he or she will say that it is a subject that can give a lot of pleasure. But that doesn't necessarily imply that society as a whole should indulge mathematicians. We might enjoy it, but we can't just go off to some grant body and say 'please give us lots of money because that will make us happy.' That's not enough.

I will discuss this point in due course. But there is another question that I need to consider first, which is why the game of mathematics has anything at all to do with the real world. How can mathematics make any difference? How is it that people employ mathematicians to do various tasks, if all they are doing is changing one set of symbols into another set of symbols? This is sometimes held to be very mysterious, but I think it is not mysterious at all. I have just discussed why mathematics is a difficult thing to do, but if we look at the experience of mathematicians, a more positive picture emerges. Very often, we want to know whether a certain statement is true. A lot of people try their hands at the problem, and only after some time does somebody come up with a solution. It can seem almost miraculous when that happens, but the point is that it is not impossible. And an important reason for its not being impossible is that the symbols we are playing with are not meaningless. The meanings we attach to them play a huge part in how we decide to manipulate them, as do our powers of visualization. In a sense, a Platonist part of us comes to the rescue (even if we claim to be formalists). Thus, Platonism is a good description of our psychology, even if there is no independently existing abstract realm.

But how does mathematics affect the real world? Let me give you a rough indication of what I would regard as the link between mathematics and science. This is oversimplified; I know philosophers of science would say that there is no such thing as a raw observation, that all observations are colored by theory. But let me oversimplify. You start with some observations. You go out and look at the world, and maybe spot some patterns of behavior. Those lead you to formulate a theory to the effect that the pattern actually describes something more general that you could expect to see again.

Once you have formulated a theory, you must usually make some simplifying assumptions. For example, if you are describing the motion of a vehicle, you might assume there is no wind resistance operating, and so forth. So when you do science, you are really talking about a model of the world, rather than the real, messy world in all its endless complexity. That's where the mathematician comes in. If you describe your model in a precise way, he or she can make some calculations within the model and use them to make predictions. That is extremely useful, because often the predictions are far from obvious and are obtained from the model only after a lot of hard work. To give a very well known example of this process, Johannes Kepler (who was more mathematician than astronomer), using astronomical observations of the positions of heavenly bodies, was led to formulate his famous three laws of planetary motion. Upon scrutinizing these, Isaac Newton suggested that their cause was gravity — the same gravity of which we are so well aware in daily life — and that gravity is a force that decreases with distance from the attracting body in accordance with the inverse square law.

Now, it would be completely wrong to say that Newton passed directly from observations to his inverse square law. He actually did some highly innovative mathematics to work out that the inverse square law would cause the planets, which he assumed to obey his own second law of motion, to move in elliptical orbits with a focus at the sun, as required by Kepler's first law. But if you want to apply Newton's inverse square law to predict what will actually happen in the sky in the future, you will often have to do some simplifications, because without them you end up with differential equations that are known not to be soluble in an exact way. You might assume, for example, that the sun is fixed in space, thus ignoring its motion around the Milky Way galaxy, and you might restrict the calculation to a single planet, ignoring the presence of other planets that are exerting gravitational forces on the planet you are interested in. That gives you a simplified model. You might even think of the sun and the planet as single points with no extension to them. Calculations done under these simplifying assumptions lead to predictions of where the planet, or perhaps a comet, will be at any given time in the future. As we all know from recent events, these predictions are often extremely accurate. Eclipses can be predicted, many years in advance, to within the minute — something that you wouldn't be able to guess unless you were prepared to sit down and do some serious calculation.

Let me offer a few more well known cases in which mathematicians (or scientists doing mathematics) have looked at models and made highly unexpected predictions that have later been verified experimentally. James Clerk Maxwell formulated his famous set of equations by thinking about the relationship between electricity and magnetism. From these equations it follows that the speed of light *in vacuo* must be constant and that electromagnetic waves other than light waves must exist. Sure enough, a few years later Heinrich Hertz actually discovered those new waves, radio waves, which have turned out to have extensive and momentous applications. And the speed of light was indeed shown to be constant by Albert Michelson and Edward Morley, a finding that was an important spur to Einstein's theory of special relativity.

The Englishman John Couch Adams and the Frenchman Urbain Le Verrier, starting from Newton's laws and the observation that the orbit of Uranus was not quite as predicted, concluded that the disagreement must be attributable to the influence of another, as yet unknown planet. They were able to calculate where that new planet ought to be and where to look for it in the sky. To cut a complicated story short, eventually the German astronomer Johann Galle did look in the right place and found the planet, Neptune, just as the mathematics had predicted. At the time, the idea that you could abstractly predict a planet's position ahead of observation was pretty controversial. It took several years for people to take it seriously.

Albert Einstein's theory of general relativity, the modern theory of gravity, predicts the existence of black holes. I don't know whether anyone has actually observed them or whether there is now convincing evidence for a black hole in any particular place, but everybody believes that they exist.[1]

John Pople got the Nobel Prize in Chemistry in 1998, but he was really a mathematician. His work has greatly improved the methods for solving the quantum-mechanical equations that are needed for computer simulations of systems dealing with lots of particles. With his results, one can look at very complicated collections of particles and actually simulate the chemistry, which was unthinkable before his work. This opens the possibility for people working on computers — which basically have mathematics underlying them — to make predictions about chemistry. Chemists can then go off to the laboratory, put the predicted compounds together in their test tubes, and verify the predictions.

Lastly, there have been many examples of elementary particles that were predicted by theory and discovered only years later, in experiments using particle accelerators.[2] The investigators have to know what to look for, and it is elementary particle theory, which is heavily based on mathematics, that tells them.

The title of this talk is deliberately somewhat paradoxical: 'What Can Pure Mathematics Offer to Society?' You might say that everything I have discussed so far is really about what people would think of as applied mathematics, almost by definition, because it has an effect on science. So what about real, pure mathematics?

Let me give a few examples of things that I think everybody would agree are pure mathematics. A classic theorem of pure mathematics is the prime number theorem. If you look at prime numbers, one of the first things you notice is that it is very difficult to spot a pattern, or at least a good enough pattern to allow you to guess what the next prime number is going to be. Of course, you know the next prime number is going to be odd, and a few other basic facts of a similar kind, but, given the knowledge that in passing from one prime to the next you have jumped by 2 and then 4 and then 2 and then 8 and then 12 and then 2 and then 2 and then 4 again, can you predict the jump to the next prime?

If you check each whole number in turn, making a note of whether it is a prime or not, you will not spot any obvious system. However, if you stand back and look at the primes from a distance, so to speak, then they

---

[1]Evidence for the existence of gigantic black holes in the hearts of many galaxies has multiplied in the years since this lecture was given. — *Ed.*
[2]The first such particles were actually discovered using cosmic rays from natural particle acceleration processes, whose nature is still something of a puzzle. — *Ed.*

come to seem much more regular. For instance, you never get a huge clump of primes followed by a huge interval where there are almost none. The primes are rather evenly spread out, and their randomness is apparent only if you look at the fine detail. In a large interval of numbers you get a certain density of primes, and this density gradually goes down as the numbers get larger, in a rather nice way that can be described as follows.

If you want to know how many primes there are up to number N, a very good approximation turns out to be N divided by the natural logarithm of N, which means the logarithm to base $e = 2.71828\ldots$. You can even improve this approximation with a slightly more elaborate expression. Most people would agree that this approximate density constitutes one of the best theorems of mathematics, known as the prime number theorem. It was proved at the end of the nineteenth century by Jacques Hadamard and independently by C. Jean de la Vallee-Poussin. The suggestion that it might be true goes back to Adrien-Marie Legendre and Carl Friedrich Gauss; even in those pre-computer days, they had enough of a feel for the primes to guess that the above formula was probably the right answer. Part of the reason that this is considered one of the classic theorems of mathematics is that its proof is very interesting, but I haven't got time to talk about that.

There's also L.E. Jan Brouwer's fixed point theorem, which considers what happens to a disc when you send it continuously into itself. Imagine a disc made of rubber that you can stretch however you like, as long as you do not tear it. For example, you can stretch it out a bit and fold it round into a sausage shape and put it back down; or you can even, a bit more drastically, scrunch it up, drop it on the floor and stamp on it so that it folds all over itself. The theorem states that however you stretch it, there will be at least one point on the rubber disc that ends up exactly where it started, assuming only that all of the rubber is inside the boundary of the disc as it was before it was distorted. This is quite a nice fact, and, again, it is unexpected. It is also a good example of the kind of game that we eccentric mathematicians like to play: We are thinking about scrunching up bits of rubber while scientists get on with more practical research that helps to drive technological progress.

Let me add one more example: the four-color theorem. This one started out in the late nineteenth century as the famous four-color problem[3] and became a theorem only in the 1970s. The proof was rather controversial at

---

[3]The relevant conjecture is attributable to Augustus de Morgan. — *Ed.*

the time, because it relied on a huge verification that was carried out by a computer. Nevertheless, it is now generally accepted.

Suppose you have a map of a few contiguous countries, which you would like to color in such a way that no two countries sharing a border will have the same color. It was conjectured in the late nineteenth century that at most four colors are needed. Very soon after, it was proved that one can always color the map with no more than five colors. While experience showed that you could also do it with just four colors, this turned out to be much harder to prove. Various false proofs of the four-color theorem were published over the years, including one that was accepted for quite a few years before its incorrectness was noticed, but eventually the issue settled down as one of the famous unsolved problems of mathematics.

This problem can be reformulated as follows. Suppose you put a node in each country and join the nodes of any two countries that share a border with a straight line, or *edge*. If, instead of coloring whole countries, you color just the nodes, then you want no two nodes linked by an edge to have the same color. Mathematicians call such a system of nodes and edges a graph. The problem of how many colors are needed for the nodes is equivalent to the problem of how many colors are required for the map, so let's get rid of the original map and look at the associated graph instead. This is called a 'planar graph,' because, if the locations of the nodes are chosen judiciously, the graph can be drawn in a plane without any two of its edges crossing over each other. Now, if you try to color a planar graph with only red, green and blue nodes, you quickly get into trouble. For example, imagine four nodes with one lying inside the triangle formed by the other three. Since the three outer nodes can all be joined to the inner one (a country can be surrounded by three countries), we need an extra color: Red, blue and green are not enough. Showing that four colors *are* enough is more complicated. But in fact no graph, however complicated, requires more than four colors for its nodes.

Although it may sound hard to believe, I first chose these three theorems as examples of pure mathematics, and then I thought, why not see whether they have any connections to the world beyond mathematics. They all did. The standard and very good way of thinking about the prime number theorem is to examine what mathematicians call the *zeta function*, which, unlike the simple functions taught in basic high school math courses, is a function of complex numbers. Many qualities of prime numbers are equivalent to properties of this mysterious zeta function. For example, an important refinement of the prime number theorem would follow from the famous hypothesis by G.F. Bernhard Riemann that the zeros of the zeta function — those locations where this function takes on a zero value, of which there must be an infinite number — all lie on a particular line in the plane of complex numbers. People have searched for these zeros, and all the known ones do lie on this particular line.[4] It is also known theoretically that a positive proportion of the zeros lie on the critical line. Mathematicians have made statistical analyses of the locations of the zeros, which seem to be rather randomly distributed along the line. But what does 'randomly distributed' mean, here or in any other problem? In general, it means that particular quantities of interest are distributed according to a specific probability distribution. Astonishingly, the probability distribution of the zeta function zeros seems to be exactly the same as a distribution that appears in the world of 'quantum chaos.'

In the last decades, the popular notion of chaos has become well defined mathematically, and physicists have discovered many systems of interest to them whose motion or changes are 'chaotic' in this precise sense. Predictions of future states of a system subject to chaos become well-nigh impossible. For example, the development of weather systems is chaotic, which is why the weather cannot be predicted with any certainty more than a few days in advance. Physicists have also studied the analogue of chaos in the quantum world — quantum chaos. It's hard to imagine anything more fashionable-sounding at present than the juxtaposition of 'quantum' and 'chaos.' And now it turns out that the distribution of the zeros of the zeta function along the notorious line in the complex plane seems to accord with a probability distribution appearing in quantum chaos dynamics. This is a most unexpected link between number theory, with its questions about the prime numbers — the purest of pure mathematics — and a phenomenon that fascinates physicists and can be studied in the laboratory. There is a lot of excitement about this link, and many people are studying it.

As for Brouwer's fixed-point theorem, this was never just an amusement; people had reasons for wanting to prove it well before they actually did so. One has to do with solving differential equations (which scientists do all the time), since it was discovered at an early stage that many of them cannot be solved using exact formulae. You cannot always write down some sort of explicit expression and verify that it satisfies the differential equation.

---

[4]Excluding the so called 'trivial' zeros such as -2, -4, etc., which are not on the said line, but this does not affect the momentous consequences emanating from the Riemann hypothesis. — *Ed.*

So attention turned to establishing that solutions exist for this or that differential equation of interest, even if the formulae for those solutions cannot be displayed. But how do you prove that something exists without actually describing it in detail?

Well, there are a number of methods for doing this, and one of them uses the fixed-point theorem, which tells you that under any transformation there must be at least one point that goes to itself, without telling you which point. There are clever means to reduce certain differential equations to a statement involving this fixed-point theorem, from which it may be concluded that solutions to these equations exist. This method, in the cases for which it works, has very natural connections to physics and to other disciplines like economics. To give a striking example, John Nash, who was awarded the Nobel Prize in Economics, worked in game theory, a branch of mathematics that has found notable applications in economics. One of his most important results, his demonstration that certain games have equilibrium states, used Brouwer's fixed-point theorem.

The four-color problem led to graph theory, which has a close relationship to computer science. You can't really be a theoretical computer scientist these days without knowing quite a bit about graph theory. Rather than go into details about that, let me convince you that graph theory has extremely practical applications. Imagine the following problem, one that arises in the real world. You want to apportion times to various people for various activities while avoiding clashes. For example, perhaps you have a group of candidates who are to take certain examinations. We cannot have examinations in two papers at the same time if there is a candidate taking both papers. How do we get through the examinations in the shortest time possible?

We can convert this into a graph-coloring problem as follows. For each paper we draw a node, and we join two nodes with an edge if the papers they represent cannot be at the same time. If we can then find a way of coloring the nodes with four colors — say, red, blue, green and yellow — in such a way that no two nodes with the same color are joined, then we can have all the red-node papers at the same time, all the blue-node papers at the same time, and so forth, and there will not be any clashes. So another problem that looks like a mere curiosity, graph coloring, turns out to be more or less equivalent to a very practical question.

Now I would like to discuss a few unsolved problems. I may have convinced you that a lot of the mathematics that has already been done has been useful, sometimes unexpectedly so. But perhaps that is the end of it.

Who knows what will happen in the future? Maybe the world has already got what it needs from mathematics. So here is an unsolved problem I like very much that is closely related to other problems I have worked on. Consider the following sequence of whole numbers: 5, 11, 17, 23, 29. I have just jumped up by 6 each time to get the next number. In the sequence 7, 19, 31, 43, I have jumped up each time by 12. And in 7, 37, 67, 97, 127, 157, the jump between successive numbers is 30. All these are prime numbers. But had I tried to continue the first sequence by adding 6 once more, I would have got 35, which is not a prime. Similarly, adding 12 to the last number in my second sequence would yield 55 — not a prime; and the next member in my third sequence is 187 — again, not a prime. So I couldn't go any further with any of these sequences.

Once you see this, you might ask yourself some natural questions; for example, what is the longest sequence of primes you could get by adding the same number each time? Is there a longest sequence? Maybe you could do it for as long as you like. There is a group in Australia that likes to explore this question with the help of computers; they have found sequences of over twenty primes with the same jump from each one to the next. But pure mathematicians don't really like that approach. You can do twenty primes, but could you do a million primes? No computer in the world, now or in the foreseeable future, can tell you. It has been conjectured that it should be possible to find arbitrarily long arithmetic progressions (sequences like those above) consisting of prime numbers, but nobody knows whether this is true or not.[5] To verify it, you would have to do some proper work to produce a 'proof.' A related question would be: Can you find infinitely many different examples of arithmetic progressions consisting of four primes each? Even that is not known. Again, that's a pretty pure-mathematical question to ask. But I can try to play the same game as I did before, by giving a few problems that look rather pure and then showing that they are closely related to much more practical questions. I will try to draw some general conclusions from all this.

One of these problems, a more geometrical question, is the Kakeya problem. The original question asked by Soichi Kakeya referred to a needle, but you can think of it as a worm, a rather rigid worm. You try to find an apartment for this worm with as little floor space as possible. However, this worm is mildly claustrophobic; it likes to be able to rotate itself through

---

[5]This conjecture was proved in 2004 (after this lecture was given) by Ben Green and Terence Tao.

360 degrees. The problem is to determine how small an area the apartment can have while still allowing this rotation. Initially, it was thought that the apartment would need to be in the shape of a kind of triangle with sides that curve inward, leaving just enough space for the worm to reach into one vertex and so manage to move its other end past the constriction of the opposite side. Eventually, the startling conclusion was reached that you can actually make this apartment as small as you like. If space is at a premium, a millionth of a square foot would be enough, as long as the worm is very thin.

A related question — in fact, roughly the same one — is the following. Suppose I take a triangle and cut up its base into a certain number of pieces; say, eight. Each piece becomes the base of a new skinny triangle whose apex is that of the original triangle. Now I want to slide those pieces about in such a way that they overlap, so that the total area covered by those pieces is as small as possible. How should I do it? The answer, it turns out, is to take the pieces in pairs, slide them together to get a bit of overlap in each pair, and then to take pairs of overlapping pairs and make those overlap, and finally to take the two clumps of four triangles and make those overlap. The result is a sort of tree-like shape. Now repeat the experiment, but cut the triangle into more pieces. As long as you cut it up into enough pieces to start with, you can make the area of the tree shape you end up with as small as you like. However, to achieve a small final area, you need to cut the original triangle up into a very large number of pieces. For example, to make it a twentieth of the original area, you need to cut the triangle up into about a million pieces. And if you want to go down to 1% of the original area, you need a really huge number of pieces.

So, roughly speaking, how efficiently can you make these triangles overlap? What's difficult about this is that they are all pointing in different directions, so that getting them to overlap is hard. But it is known that you can get the final area to be as small as you like by cutting up the triangle into enough pieces. The unsolved problem comes when you look at a higher-dimensional generalization of this. Suppose that instead of cutting a triangle's base into a number of pieces and sliding them about, I take a square pyramid and cut its base into several pieces, thus making a set of skinny square pyramids which I can then slide about. Now, by making them overlap as much as possible, I want the covered volume to be as small as possible. Can I do better in three dimensions than I can in two? If I want to make the total volume a twentieth of what it was before the cut-up, do I still need about a million pieces, or can I do it in some clever way with fewer? Can I slide things about in three dimensions more efficiently than in two? Or, conversely, if I have the same number of pieces as for the triangle, can I make the volume go down further in three dimensions than the area would have gone down in two? As I said, the answer is not known.

Maybe pure mathematicians enjoy thinking about that sort of thing, but it is not obvious that they should be paid to do so. Actually, however, a lot of the things I have talked about have links to each other, not always in expected ways. Thinking about arithmetic progressions of prime numbers led to more general questions about sets and arithmetic progressions. I don't want to say too much about that; however, a famous theorem of Endre Szemerédi was proved in a different way by Hillel Furstenberg (who happens to be in this room) using ergodic theory, a branch of mathematics that has a very direct connection with physics. That is, a pure question about prime numbers yielded a two-stage link, somewhat indirect but real, between that question and parts of mathematics that have direct connections with physics. I myself thought about trying to find yet another argument for Szemerédi's theorem, and was thereby led to improve another theorem, called the Balog–Szemerédi theorem; and from this resulted a third proof of Szemerédi's theorem. Again, if you are not a mathematician, you might wonder why one should bother proving something when everybody already knows it's true. Well, there are very good reasons for doing it.

This very case provides an example, since Jean Bourgain — by modifying an argument that I had used in thinking about something quite different — was then able to use it to obtain the best known results about the Kakeya problem in high dimensions. The Kakeya problem was known to be equivalent, or at least very closely related, to questions about the behavior of certain non-linear partial differential equations, which for several reasons are the most interesting kinds of differential equations. They describe a number of physical phenomena (e.g., hydrodynamics, relativistic gravitation, the strong interaction between elementary particles, etc. — *Ed.*), and they are much harder to analyze than linear differential equations. Physics is thus full of non-linear partial differential equations that people would like to understand, and the Kakeya problem contributes to that understanding. Bourgain had previously shown that the Kakeya problem is intimately connected with a conjecture by Hugh Montgomery, which itself is intimately connected with the distribution of the zeros in the zeta function. I have already told you that the latter is intimately connected with the behavior of prime numbers, and that you need to know about its behavior if you want to think about arithmetic progressions of prime numbers.

So there is a big circle of problems with close connections, and from one area of mathematics you can arrive very quickly at another completely different area. That doesn't necessarily mean that if a mathematician thinks about prime numbers, then somebody else who is a physicist will come along and say: 'Ah! That's exactly the result I needed.' What it does mean is that thinking about many different problems contributes to a general mathematical culture which scientists can then come in and use. It is as simple as that: If you try to inhibit work in the 'useless' areas of mathematics, the useful ones will suffer, too.

I would go so far as to say that at any one time, most mathematicians are working on problems that are unlikely to be of use to scientists or to anybody else. It might even be true that most of their papers are hardly read even by other mathematicians, because people would rather be getting on with their own research. However, this is not an argument for trying to restrict the activity of mathematicians, because, as I have been saying, mathematics is an incredibly interconnected subject. If you try to restrict some mathematicians working in 'not useful' areas, you can never be quite sure that, somewhere down the chain of connections from their work, there might not be mathematicians who are doing something that society has a direct interest in, and whose work would thereby be the poorer.

Very often, practical applications of mathematics come when you least expect them. The example that everybody gives is the RSA public-key encryption system of Ronald L. Rivest, Adi Shamir and Leonard M. Adelman. G.H. Hardy famously said that he worked on number theory because of its beauty, and that it had no applications then or in the foreseeable future. Not so long after he made that prediction, number theory was applied directly to produce the RSA system. Even when that was published, I think people wouldn't have predicted that it would later become extremely useful for the Internet. Nowadays, if you are able to make a safe online credit-card transaction, you should be very grateful to the pure mathematicians who, already hundreds of years ago, developed the number theory that lies at the basis of RSA.

Another very important point to make is the following. Even though mathematicians themselves are not trying to be useful in any sense at all — I think most of them do mathematics because they love the subject, they are curious about things mathematical, and they may also want to get on in their careers — nevertheless, the collective effect of their work is very different from the sum of all their individual motivations. Of course, this phenomenon occurs in other areas as well. A basic premise of capitalist economics is that even if people are working largely for their own self-interest, this can bring economic benefits to society as a whole. That is, individual greed and selfishness can work to the common good. Maybe they don't always, but they can.

One might think of the activity of pure mathematicians as a 'data resource.' You have a whole lot of mathematicians, with lots of information in their heads, but they all have different information. It is all highly linked through contacts with other mathematicians, the Internet and things like that. Now, let's think of other examples of data resources, such as libraries, encyclopedias, the Internet or human brains. If you were to try to design them more efficiently by keeping the useful parts and getting rid of the less useful ones, the outcome, in each case, would be disastrous. If you try to improve a library by throwing away all the books that you think no one is ever going to consult, you are bound to get it wrong. The useful thing about a library, or an encyclopedia, or the Internet, is that although it contains masses of useless information, you can find in it what you want, whenever you want it. The more you can do that, the more useful it is. That applies to the mathematics community in general. This is a somewhat abstract argument for why mathematics is useful.

Let me return to some actual mathematics, although I shall not do this in any detail. I want to give one example of an area of mathematics in which many mathematicians, and not just pure mathematicians, are interested at the moment. I am interested in it very much from the sidelines, but it is something that, I am convinced, is bound to be extremely useful over the next century in unpredictable ways. I've let myself off the hook here by not saying why it is going to be useful; it just seems obvious to me that it will be. This is statistical physics, which can be viewed as a branch both of mathematics and of physics.

I will try to describe it by looking at one example, the 'Ising model,' named after the late physicist Ernst Ising. Imagine some particles arranged on a grid, with pluses or minuses representing the sense of their spins. I don't myself really understand what the spin of a particle is, but in order to follow what is going on, all we need to understand is that within the model there is a tendency of particles, when they are next to each other, to want to align their spins. So if one particle is a plus, then the one next to it wants to be a plus as well. When you have a plus next to a minus, there is a cost associated with it; as it were, the system feels a little bit uncomfortable. More precisely, the more edges there are that link sites with opposite signs, the higher the energy associated with the system.

Now, let's assign a likelihood to the whole system, so that the more edges like this you have — that is, the more instances you may have of two particles with opposite spins next to each other — the less likely you are to find the system in such a state. For people with mathematical training, what we are doing is associating a probability with the system (namely, the probability of an equilibrium state — *Ed.*). We make it proportional to e (the base of the natural logarithms) raised to minus B (a constant related to the temperature) times H, the number of the aforementioned edges (a stand-in for the system's energy — *Ed.*). Now, all you really need to know about this is that when H is large, the probability becomes small, while when H is small, the probability is large. So a state of our system with a large H (lots of those edges) is rather unlikely, while one with a small H is more likely.

An example of a likely state is one with all the pluses to the right and all the minuses to the left of a straight line; this would have a small number of plus/minus edges. But such an arrangement is not absolutely essential; there are quite a few ways to arrange a likely state. You could have a certain number of pluses grouped together and then some edges connecting to a region that predominantly has minuses. Such a state, individually, is a little less likely than the first, because the pluses have moved somewhat into the minus area and vice versa, so we have rather more edges than are strictly necessary. However, there are many more possible arrangements of the latter kind, so overall it may be much more likely that the system will be in a state of the latter kind than in the perfectly ordered state I first described.

Now, the number B here is essentially the inverse of the temperature we may ascribe to our system. If B is zero (an arbitrarily high temperature), then effectively you are not assigning any cost at all to those edges. In this case, any arrangement of pluses and minuses is equally likely. But if B is large (a low temperature), that makes the cost of an edge relatively high, which reduces the probability of a state with many edges. Therefore, at a low temperature the system is much more likely to arrange itself in some sort of relatively ordered state, with not too many edges joining large regions full of pluses with other regions full of minuses. This conclusion corresponds to the physical experience of what happens when you heat a spin system like the one we are conjecturing. When it's still cold, the forces that are trying to align the spins with each other dominate, so the spins all do align, and the system looks rather ordered. But as the system is heated past a certain temperature (the famous Curie temperature — *Ed.*), it suddenly becomes much more disordered. This is an example of what physicists call a phase transition, in our case from order to disorder.

I have oversimplified somewhat. Nevertheless, these phase transitions occur in a huge number of models. Very often when you have 'local' probabilistic interactions, they result in a rather mysterious phase transition. As with the Ising model, when you gradually increase some parameter, the system suddenly changes from an ordered to a disordered state. Physicists have applied all sorts of arguments to these phase transitions, and they can make lots of predictions about these models. However, mathematicians feel slightly suspicious about these arguments, because they do not meet the standards of rigor that a pure mathematician would require. I think most mathematicians would summarize what physicists have done on this subject by saying that it is very clever, that the mathematics behind it is manifestly wrong, and that, apart from a few exceptional cases, the non-rigorous arguments of physicists have produced the right answers (which are, moreover, answers that would be impossible to guess).

Thus, mathematicians concede that many results in the theory of phase transitions are almost certainly correct, even if the justifications for them are shaky. This fascinates mathematicians; they want to put the physics on a higher level of rigor. It is not that we are neurotic and unwilling to believe the results unless we have dotted the i's and crossed the t's. Most mathematicians would say that they do believe the results, but that something interesting is going on that deserves to be investigated further. The physicists have arguments that lead to the right answers, and nobody quite understands why they succeed; yet it seems completely inconceivable that this is an accident. The whole history of mathematics suggests that in a situation like this, if and when you finally do understand what is going on, new phenomena turn up. You don't just end up giving a pure mathematician's guarantee that what people knew all along is indeed correct; you get more than that.

The progress to be made on these questions will surely be very significant. There are a great many systems where small units are linked to each other on a small scale, and these linkages have large-scale effects. The brain, for example, has lots of neurons next to each other that all fire in a staccato pattern, and it seems clear that phase transitions must take place there. Perhaps one such transition has caused me suddenly to change my mind and turn around and look over here. But this is not at all a well-understood process.

Another application might lie in the design of computer architecture. Here, you have lots of computing modules that communicate with each other, but how are you to design the whole net of them so that it does what you want efficiently? Questions like these are very statistical-physics-like in character.

Let's look at another biological application. Contrast the behavior of a flock of birds with that of a swarm of mosquitoes. In both cases, the choices that the animals make when flying are governed by what a few neighbors are doing. But in the case of the birds, the tendency to fly similarly to one's neighbors is stronger, with the result that flocks look far more orderly than swarms. This has been modeled on computers. As you decrease the strength of the interactions, there is a phase transition, and flocks lose their order and turn into swarms.

Yet one more application: crowd psychology. How does a football chant start? I have never understood that. People can't hear what is going on around the stadium; they just hear what their neighbors are doing. Suddenly they are all singing the same thing. Some sort of phase transition has taken place there. It goes without saying that there are more important examples of crowd psychology than this one.

Let me end with something very fanciful that takes me back to the formalist view of mathematics. If you have just a very few replacement rules for getting from one string of symbols to another, it's quite easy to analyze exactly what you can derive from it. To see whether you can get from A to B, you apply the rules, and if you don't get to B, then you know you can't get to B, and that is the end of the story. On the other hand, if you have too many replacement rules, you can more or less get from anything to anything; then the game is boring because it's just too easy. Somehow, real mathematics seems to lie at some intermediate level; there's something like a phase transition at a point where you have just enough rules to keep you interested, but not quite enough to make you uninterested again. Maybe understanding a bit more of this issue could actually shed light on mathematics itself. I don't offer this as a justification for pure mathematics, but it might be a welcome spin-off.

Prof. W. Timothy Gowers delivered the Albert Einstein Memorial Lecture in 2001.

General Covariance and the Passive Equations of Physics

Shlomo Sternberg

1. What Do I Mean by 'Passive Equations'?

By the passive equations of physics, I mean those equations that describe the motion of a small object in the presence of a force field, where we ignore the effect produced by this small object.

For example, Newton's laws say that any two objects attract one another. But if we study the motion of a ball or a rocket in the gravitational field of the earth, we ignore the tiny effect that the ball or rocket has on the motion of the earth.

If we have a small charged particle in an electromagnetic field, the Lorentz equations describe the motion of the particle when we ignore the field produced by the motion of the particle itself.

To explain what I mean by general covariance will take the whole lecture.

2. The Sources of This Lecture

The first source of my lecture is a late paper by Albert Einstein, Leopold Infeld and Banesh Hoffman entitled 'The Gravitational Equations and the Problem of Motion,' published in the *Annals of Mathematics*, 39 (1938). It opens with the following words:

> In this paper we investigate the fundamentally simple question of the extent to which the relativistic equations of gravitation determine the motion of ponderable bodies.

It will take a bit of effort to explain what this 'fundamentally simple question' is. I should comment that the Einstein–Infeld–Hoffman paper is technically difficult to read, because it was written before the appropriate mathematical language — the theory of generalized functions — was developed. The person who extracted the key idea from this paper in modern mathematical language was J.M. Souriau, who applied the

Einstein–Infeld–Hoffman method to determine the equations of motion of a spinning charged particle in an electromagnetic field. His paper, 'Modèle de particule à spin dans le champ électromagnétique et gravitationnel,' appeared in *Annales de l'institut Henri Poincaré*, 20 (1974). This is the second source of my lecture.

My purpose herein is to explain how the Einstein–Infeld–Hoffman method, as formulated for spinning particles by Souriau, can be viewed as a principle for determining the passive equations of physics in a very general setting.

Figure 1. Jean Marie Souriau.

Souriau's paper is itself not an easy read. He has a wonderful but idiosyncratic mode of exposition. For example, here is the flow chart presented on page 2 of the paper (Figure 2).

3. What is the 'Fundamentally Simple Question' Posed by Einstein, Infeld and Hoffmann?

There are two fundamental principles of general relativity:

- The distribution of energy-matter determines the geometry of space time.
- A small piece of ponderable matter moves along a geodesic in the geometry determined as above.

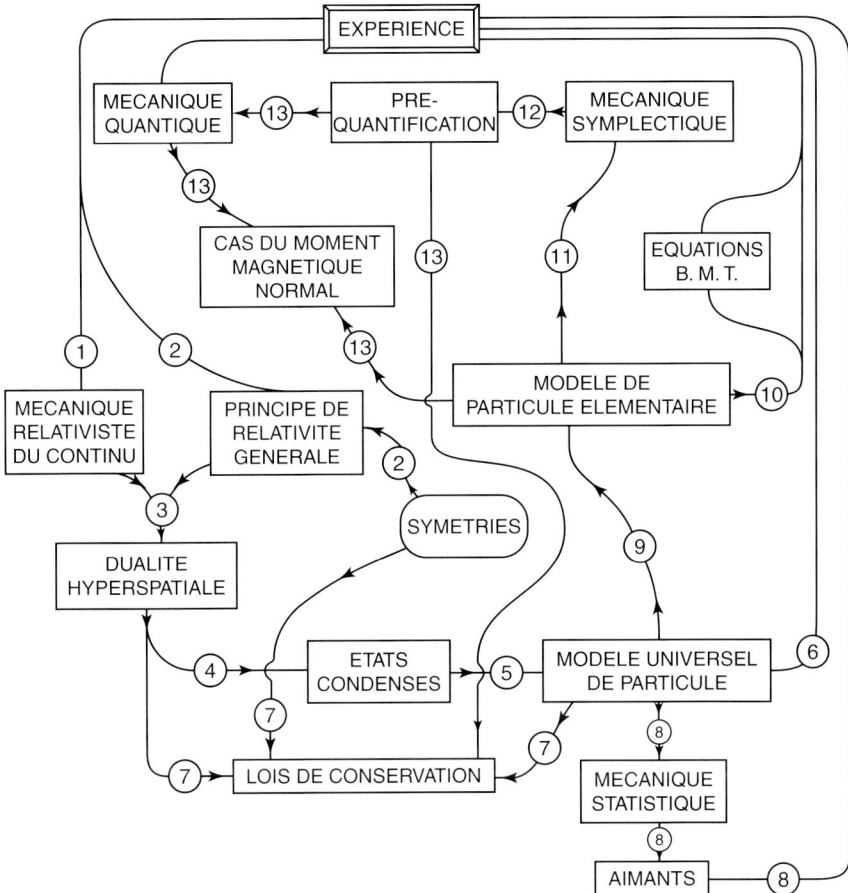

Figure 2. Flow chart, J.M. Souriau, 'Modèle de particule à spin dans le champ électromagnétique et gravitationnel,' *Annales de l'institut Henri Poincaré*, 20 (1974), p. 2.

I will spend some time explaining the meanings of the word 'geodesic.'

Many distinguished physicists thought that these were two independent principles. The point of the Einstein–Infeld–Hoffman paper was to explain how they are related.

4. Einstein's Comment on the First Principle

Referring to the impact of the work of his predecessors Heinrich Rudolf Hertz and Hendrik Antoon Lorentz, which led to the elucidation of the

Figure 3. 'People slowly accustomed themselves to the idea that the physical states of space itself were the final physical reality' — Albert Einstein (cartoon by Rea Irvin, *The New Yorker*, 1929; © Rea Irvin/The New Yorker Collection).

first principle, Einstein remarked: 'People slowly accustomed themselves to the idea that the physical states of space itself were the final physical reality' (Albert Einstein, 'The History of Field Theory,' lecture to the general public, February 3, 1929).

Figure 3 shows *The New Yorker*'s take on Einstein's comment.

5. What Is a Geodesic?

Before the papers by Einstein–Infeld–Hoffman and Souriau, there were several (equivalent) definitions of what a geodesic is. They all try to extend to more general geometries a characteristic property that straight lines have in Euclidean geometry:

- A straight line is 'the shortest distance between two points.'
- A straight line is 'self-parallel' in the sense that it always points in the same direction at all its points.

A curved line will (in general) be pointing in different directions at different points.

For example, on a sphere, the geodesics are (portions of) great circles. To see this, here is a sphere drawn using the computational software of MATLAB (Figure 4). That is, if you type the word 'sphere' in MATLAB and hit the return button, this is what you get.

Figure 5 shows a curve on the sphere, starting at the north pole. Notice that the great circles emanating from the north pole (the circles

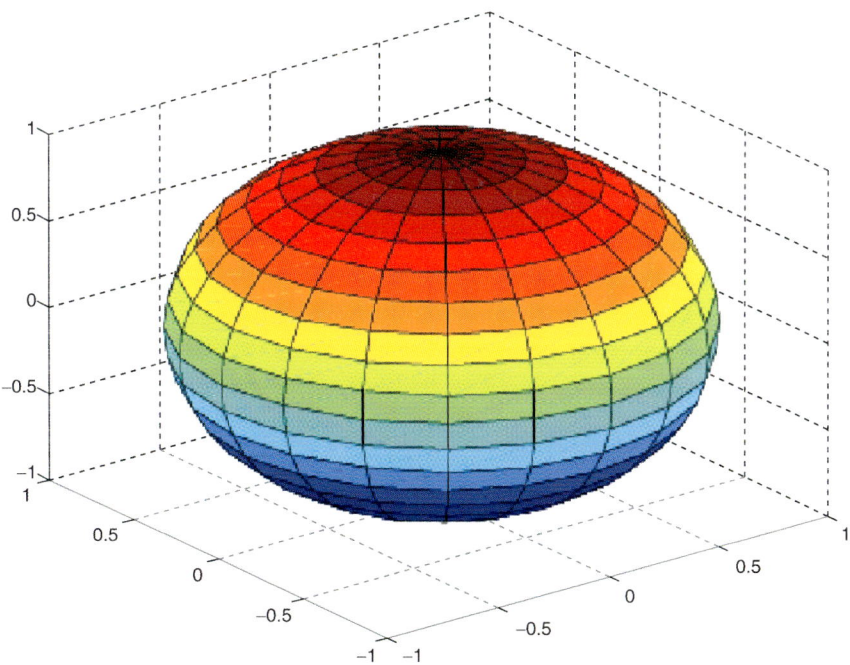

Figure 4. Sphere generated by MATLAB.

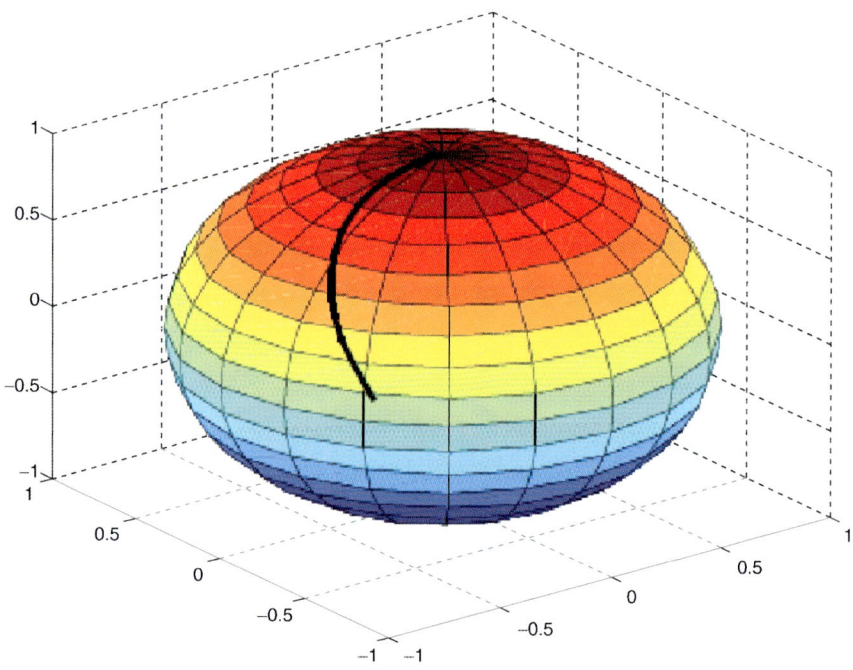

Figure 5. A curve on the sphere.

of longitude) are consistently shorter than the corresponding piece of the curve.

6. Gauss's Lemma and the Problem of Absolute Space

We can take the previous picture and rotate the point of view. This is a straightforward manipulation in MATLAB. Figure 6 shows what we get if we view the picture from the top.

Notice that from this point of view, the circles of longitude look almost like straight lines, and these lines are perpendicular to the circles of latitude. This is an illustration of a special case of what is known as Gauss's lemma, although in a sense this was anticipated by the great Islamic scientist al Bīrūnī (973–1048).

One of the philosophical problems in Newton's theory of motion was to give a meaning to the concept of 'absolute space,' in which (according to Galileo) particles subject to no force move in straight lines. The problem is that what is 'straight' in one frame of reference may not be straight in

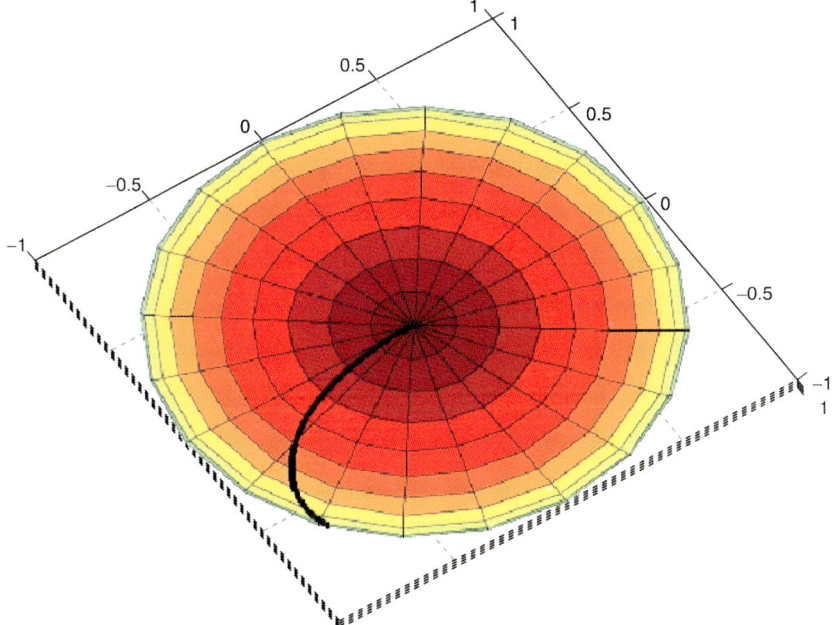

Figure 6. The sphere viewed from above (generated by MATLAB).

another. Newton's solution of this problem was theological: Absolute space existed as 'God's sensorium.' Its non-theological resolution had to await Einstein, who asserted that particles move along geodesics, and near each point there is a coordinate system (a preferred 'point of view' if you like) in which the geodesics look like straight lines.

7. The Contributions of Gauss and Riemann

The geometry of surfaces, especially the 'intrinsic' geometry of surfaces — those properties of surfaces that are independent of how they are embedded in Euclidean space — was developed by Johann Carl Friedrich Gauss (1777–1855).

But the full higher-dimensional notion of the intrinsic geometry of a possibly curved space was developed by Gauss's student Georg Friedrich Bernhard Riemann (1826–1866). The equations for geodesics as curves that locally minimize arc length play a key role in this theory. Riemann's theory of the curvature of such spaces played a key role in Einstein's theory of general relativity.

Figure 7. Abū al-Rayḥān Muḥammad b. Aḥmad al-Bīrūnī, b. 15 September 973 in Kāth, Khwārazm (now Kara-Kalpakskaya, Uzbekistan); d. 13 December 1048 in Ghazna (now Ghazni, Afghanistan).

Figure 8. Johann Carl Friedrich Gauss, b. 30 April 1777 in Brunswick, Duchy of Brunswick (now Germany); d. 23 February 1855 in Göttingen, Hanover (now Germany).

Figure 9. Georg Friedrich Bernhard Riemann, b. 17 September 1826 in Breselenz, Hanover (now Germany); d. 20 July 1866 in Selasca, Italy.

8. Parallelism along Curves, Geodesics

Can we attach a meaning to the assertion that two vectors tangent to the sphere at two different points, p and q, are parallel? The answer to this question is 'no.' However, it *does* make sense if we join p to q by a curve.

Let **c** be a curve on the sphere, starting at p and ending at q. Place the sphere on a plane so that it just touches the plane at p. If u is a vector tangent to the sphere at p, we can also think of u as being a vector U in the plane, since this plane is tangent to the sphere at p. Now, roll the sphere on the plane along the curve **c**. This will give us a curve **C** in the plane, and at the end of this process we end up with the point q touching the plane. A tangent vector v at q can be thought of as being a vector V in the plane. We say that u and v are 'parallel along' if the vectors U and V are parallel in the plane. This notion of parallelism depends on the choice of the curve. A different curve joining p to q will give a different criterion for when vectors at p and q are parallel.

We can now define geodesics to be self-parallel curves — curves **c** having the property that when you perform the rolling process, the curve **C** that you get in the plane is a (piece of) a straight line. For the sphere, the curves

Figure 10. Tullio Levi-Civita, b. 29 March 1873 in Padua, Italy; d. 29 December 1941 in Rome, Italy.

c that roll out to straight lines in the plane are exactly the great circles. But we can make this definition for any curve on any surface. It is then a mathematical theorem that this definition of geodesics, as curves that roll out to straight lines, coincides with the earlier definition of geodesics as curves that locally minimize arc length.

What about more general spaces such as those considered by Riemann? Here the key result is that of Tullio Levi-Civita (1873–1941), who introduced a general concept of parallelism of vectors along curves and showed that for any Riemannian manifold there is a unique such notion with certain desirable properties, and that the self-parallel curves are exactly the geodesics in Riemann's sense.

9. Return to the 'Fundamentally Simple Question' of Einstein–Infeld–Hoffman

The question is: What do the 'relativistic equations of gravitation' have to do with the equations that determine geodesics? In order to understand the Einstein–Infeld–Hoffman–Souriau answer to this question, we really do not need to know in detail what the 'relativistic equations of gravitation' are. (This would require a whole course in general relativity.) All that we need to know is something very general about the form of these equations, in particular the symmetry that is built into them. It is an amazing fact that these symmetry conditions alone determine the equations for geodesics.

For this we need to state some elementary facts about constraints imposed by symmetry.

10. Constraints Imposed by Symmetry

10.1. *The example of an equilateral triangle*

Here is an equilateral triangle.

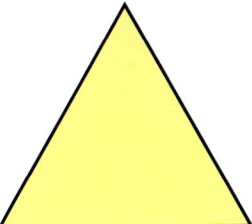

Figure 11. An equilateral triangle.

Suppose I want to attach an object, say a little disk, to one of the corners of this triangle and still preserve the symmetry. Then I must attach an identical object to all three corners.

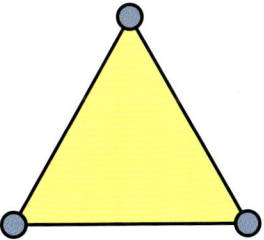

Figure 12. To preserve the symmetry, an object attached to one corner must be attached to all three corners.

If I want to place a little disk on one of the sides of the triangle and still preserve the complete symmetry, I must also place the same disk at all points (in general six of them) that I can obtain from this point by a symmetry transformation of the triangle.

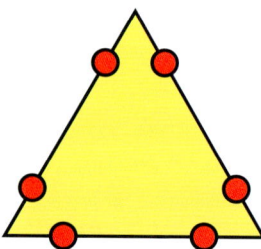

Figure 13. To preserve the symmetry, an object attached to a point on the side must be attached to all the points that can be obtained from this point by a symmetry transformation of the triangle.

Let g be the symmetry operation consisting of flipping the triangle about its vertical axis.

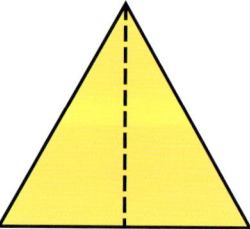

Figure 14. The vertical axis of the triangle.

If x is a point on the triangle, then gx denotes the point that is obtained from x by applying the symmetry operation g to x.

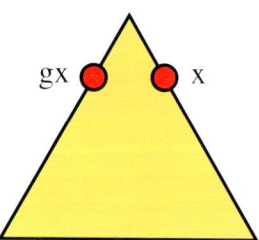

Figure 15. gx denotes the point obtained from x by applying the symmetry operation g to x.

10.2. *The concept of the orbit through a point*

There are six symmetry operations of the equilateral triangle: flipping about each of the perpendicular bisectors and rotation about the center through 0 degrees, 120 degrees and 240 degrees.

The set of all points that I obtain from x by applying all symmetry operations G to x is called the *orbit* through x and is denoted by Gx. Typically, there will be six points in an orbit.

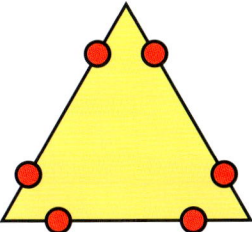

Figure 16. Six points in a typical orbit.

But there will be some orbits with three elements, for example, those through the vertices, or more generally those through points lying on any of the perpendicular bisectors.

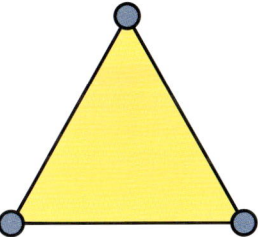

Figure 17. An orbit with three points.

And there is an orbit with only one element: 0 — the center of the triangle.

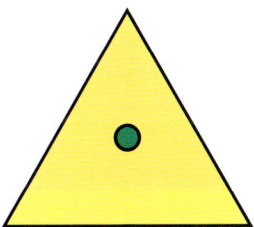

Figure 18. An orbit with only one point in the center.

10.3. *General formulation of the constraint imposed by a symmetry group*

Suppose X is a set (or object) and G a group of symmetries of X. If x is a point of X and g a symmetry in G, then we let gx denote the point of X obtained from x by applying the symmetry g. We let Gx denote the collection of all such points gx and call Gx the orbit of x under the symmetries G. Then, if F is a (say) numerical function on X, and symmetry is to be preserved, then F must take on a constant value in each orbit. Let me call this the principle of general covariance.

10.4. *Example: Spheres as orbits*

Suppose X is ordinary three-dimensional space, with a preferred point O as origin, and let G consist of all rotations about O. If x is a point different from O, then the orbit Gx is the sphere of radius r, where r is the distance from O to x. If $x = O$, then the orbit Gx consists of the single point O. Thus, the orbits are spheres centered about O, with the exception of the single orbit consisting of one point O (Figure 19). Notice that in this example, the (spherical) orbits each form a continuous manifold of points rather than a discrete collection of points, as in the preceding examples.

Our principle of general covariance (i.e., the symmetry-conserving condition) says that if F is a function, and the symmetry group G is preserved, then F must be constant on each of these spheres.

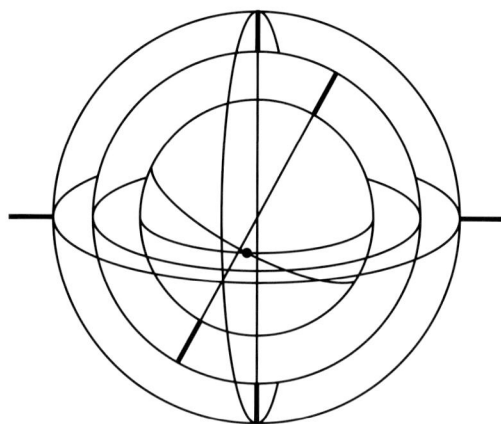

Figure 19. Orbits of the rotation group are concentric spheres.

10.5. *The infinitesimal version of the constraint*

Figure 20 pictures a function F (the varying intensity of the blue) that is constant along each curve in a family. We wish to examine the infinitesimal change in F — or, as we say, the differential (change) of F — at any point.

The constancy of F along each curve imposes the condition that the infinitesimal change ℓ of F at a point x vanishes on a tangent to the curve through x (Figure 21).

Figure 20. A function (F) that is constant along each curve in a family.

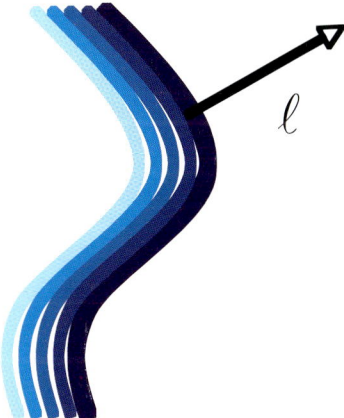

Figure 21. The infinitesimal change ℓ of F at a point x vanishes on a tangent to the curve through x.

11. The Infinitesimal Constraint Condition as a Principle of Physics

Suppose G is a group of symmetries of a space X, and suppose F is a (say) numerical function on X, and symmetry is to be preserved; then the infinitesimal change ℓ of F vanishes on tangents to the orbits.

In more symbolic language, we write this as follows: Let x be a point of X, and let $O = G \cdot x$ be the G orbit through x. Let $T_x O$ denote the tangent space to O at x. In other words, $T_x O$ denotes the collection of those infinitesimal changes in x which are tangent to the orbit.

Then if

$$dF_x = \ell$$

denotes the differential of F at x, then

$$\ell \text{ vanishes on } T_x O.$$

We now come to the punchline of today's lecture: For an appropriate choice of X and G, we can associate to certain data along a curve **c** (in technical language — a contravariant symmetric tensor field along the curve) an ℓ, that is, an object which measures infinitesimal change of a function. The condition

$$\ell \text{ vanishes on } T_x O$$

implies that the curve must be a geodesic!

So, we can formulate a general principle: A physical set of laws from this point of view is determined by

- the choice of an appropriate space X
- a group of symmetries G of X
- a function F on X invariant under the group.

Then, there are two types of equations: A *source* equation: Given ℓ, find x such that

$$dF_x = \ell \; (\mathbf{S})$$

and a necessary condition on ℓ for this to be possible, namely:

$$\ell \text{ vanishes on } T_x O. \; (\mathbf{P})$$

It then turns out that this necessary condition by itself gives a set of physical laws: the 'passive' laws of the theory.

In general relativity, for specific types of ℓ (those associated to curves), equation (**P**) determines the 'motion of ponderable matter.'

So I have to explain what X, G and ℓ are in the case of general relativity, and also say something about F.

12. Some More Geometry

I will try to give some general idea of what X, G and F are in general relativity. (ℓ will be a bit harder to explain in non-technical terms.) For this I have to go back and do a bit more geometry.

Figure 22 shows a surface in the shape of a jug.

Some things to notice: Near the base of the jug, the horizontal and the vertical curves are both curving 'in the same direction' — bending outward from the center line of the jug. At the neck of the jug, the horizontal curves are still curving outward, while the vertical curves are curving inward. We say that the Gaussian curvature is positive when the curves

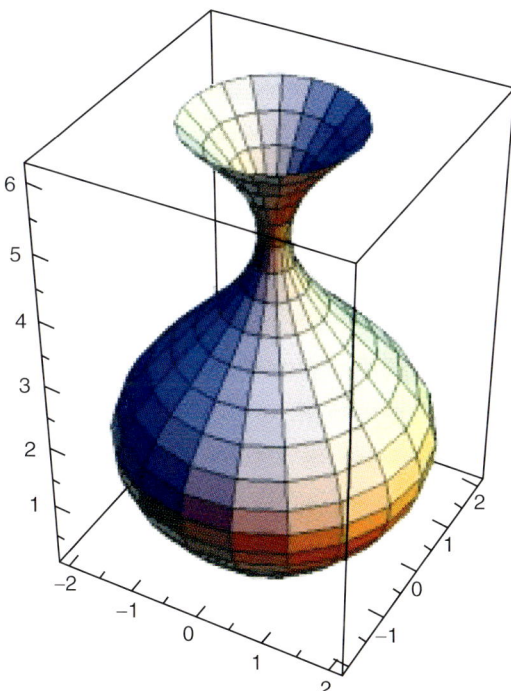

Figure 22. A surface in the shape of a jug.

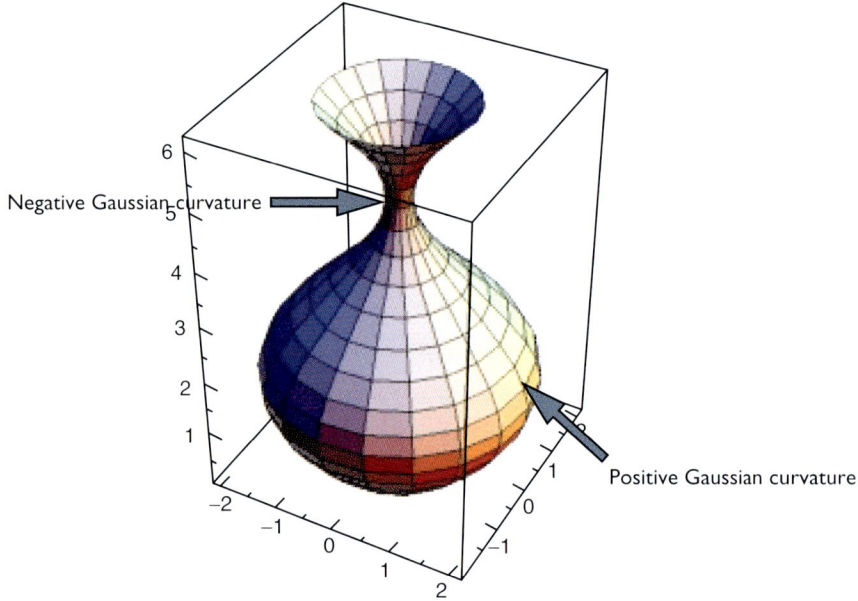

Figure 23. Positive and negative Gaussian curvature on a jug-shaped surface.

bend in the same direction, and negative when they bend in opposite directions (Figure 23).

Figure 24 shows an example of a surface whose Gaussian curvature is negative everywhere.

Of course, a sphere is an example of a surface whose Gaussian curvature is positive everywhere.

Back to the jug (Figure 22). Notice that on the jug there is a grid that measures distances for an object crawling along the surface. A bug crawling along the surface of the jug can count how many gridlines (vertical and horizontal) it crosses.

Of course, the network of gridlines on the jug is inherited from the geometry of three-dimensional space. But we can imagine a two-dimensional life form that cannot perceive the third dimension — only the grid. In fact, *Flatland: A Romance of Many Dimensions* (1884), the classic nineteenth-century novella by Edwin Abbott Abbott, describes just such a life (and hints at a possible 'fourth dimension' for us three-dimensional creatures!).

Now, sometimes two quite different surfaces in three-dimensional space can give the same (local) geometry of gridlines. For example, the local geometry of the gridlines of a cylinder is the same as the local geometry

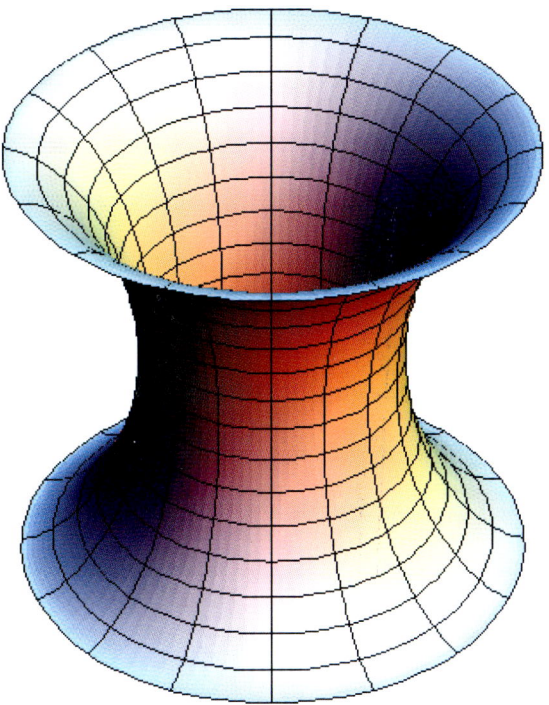

Figure 24. A surface whose Gaussian curvature is negative everywhere.

of the plane (Figure 25). We all know this from the fact that we can wrap a piece of paper around the cylinder without stretching or tearing it. We cannot wrap the piece of paper around a sphere, for example.

I use the word 'local' in the following sense: A bug walking a short distance around the cylinder (and measuring his path by the gridlines) cannot tell whether he is on a cylinder or on a plane. Of course, if he walks far enough in the right direction, he will come back to where he started, which would not be the case for the plane. But he cannot tell from a short stroll.

So we have two concepts: the 'intrinsic geometry' of a surface, determined by its gridlines, and its 'extrinsic geometry,' determined by its shape in space. The cylinder and the plane have the same (local) intrinsic geometry, but they have different shapes in space. The cylinder and the sphere have different intrinsic geometries.

I haven't given (and won't give) a precise numerical definition of the Gaussian curvature. It is a combination of how the surface is bending in various directions. But with the precise definition, it turns out that the cylinder has a Gaussian curvature of zero — as does the plane. Gauss

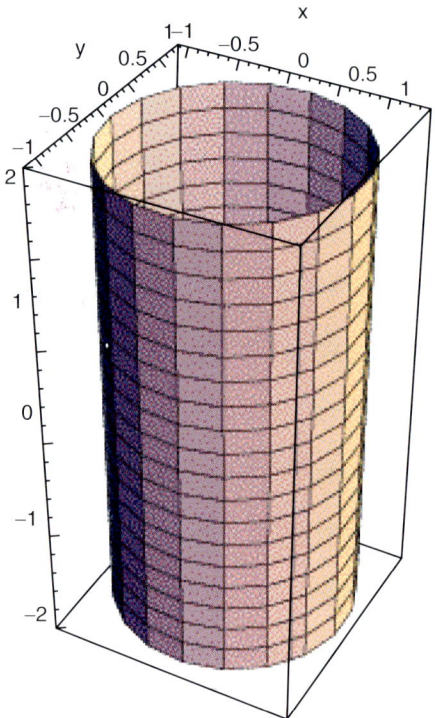

Figure 25. The cylinder and the plane.

discovered that his curvature depends only on the intrinsic geometry of the surface, not on the details of its shape in space. In fact, Gauss was so proud of this discovery that he called it the *Theorema egregium*.

This concept of intrinsic geometry was taken up by Riemann and extended to higher dimensions (as we have already mentioned), and Riemann also gave the higher-dimensional version of the Gaussian curvature.

I can now give an idea of what the space X is in general relativity: It is the space of all possible metrics. In two dimensions, the analogue would be the space of all possible choices of grids on a given surface. But of course, general relativity deals with four dimensions.

The function F is also easy to describe: The two-dimensional analogue would be the integral of the Gaussian curvature K over this surface, that is: $\int K dA$. The grid provides a notion of area as well as of length, and $F(x) = \int_S K dA$ represents a sort of average of the Gaussian curvature with respect to the notion dA of area provided by the grid.

What about the group G? Let us go back for a moment to the jug (Figure 22). Suppose it were made of some flexible material, so that we could deform it. For example, we might be able to blow it up like a balloon. Of course, when we do this, we get a different surface with a different grid, a different curvature, etc. But suppose we keep the original markings of the grid, and they get distorted along with the surface — like what happens when we blow up a balloon with some writing on it. If we use the inflated grid (from the old markings) to measure distances, we really haven't changed anything. In other words, we are letting G be the group of all smooth deformations of the surface, with their action on the grids. The function F does not change under such transformation. (Actually, for technical reasons, we want this change to take place only in a finite region. In the case of the jug, for example, we would demand that some region around the top rim be left unchanged.)

I can now repeat the punchline of today's lecture: With the above choice of X and G, we can associate to certain data along a curve **c** (in technical language — a contravariant symmetric tensor field along the curve) an ℓ, that is, an object that measures infinitesimal change of a function, and the condition

$$\ell \text{ vanishes on } T_x O$$

implies that the curve must be a geodesic!

13. The Technical Formulation for the Case of General Relativity

I shall now give a mathematically precise formulation of the main result.

Let M be a manifold. Let X denote the set of all semi-Riemannian metrics on M of a given signature. (For example, M could be space time and X the set of all Lorentzian metrics on M.) We will let G be the group of diffeomorphisms of compact support on M. (So an element of G will be a diffeomorphism of M which equals the identity outside some compact subset of M.) An element $\varphi \in G$ acts on a metric x by sending it into $(\varphi^{-1})^*x$. This gives an action of G on X.

The full tangent space $T_x O$ can be identified with the space $S_2(X)$ of smooth symmetric covariant tensor fields of degree 2. Notice that we have identified all tangent spaces with the same fixed vector space. We have *trivialized* the tangent bundle to X. We want to consider the subspace $S_2^0(X) \subset S_2(X)$ consisting of those smooth tensor fields of compact support.

The corresponding subspace of $T_x X$ will be denoted by $T_x^0 X$. It is to be thought of as those infinitesimal variations of the metric x which vanish outside some compact subset.

How should we think of $T_x O$? If u is a vector field on M, then differential geometry attaches a meaning to $D_u x$, the Lie derivative of the metric x with respect to the vector field u. It is the symmetric covariant 2-tensor whose value $D_u x(v, w)$ on two other vector fields v and w is defined as follows:

$$D_u x(v, w) = u(u, w)_x - ([u, v], w)_x - (v, [u, w])_x.$$

Here $(v, w)_x$ denotes the scalar product of v and w (a function on M) determined by the metric x.

Since G consists of diffeomorphisms of compact support, we let $T_x O$ consist of those $D_u x$ where u is a vector field of compact support. Clearly

$$T_x O \subset T_x^0 X.$$

What are the possible ℓ's? We want an ℓ to be a (continuous) linear function on $S_2^0(M)$. (This is where the theory of generalized functions comes in.) For example, suppose that τ is a smooth contravariant symmetric 2-tensor. If σ is a symmetric covariant 2-tensor of compact support, then the double contraction

$$\sigma \cdot \tau$$

is a smooth function of compact support on M. The metric x determines a volume vol_x, and we can integrate the function $\sigma \cdot \tau$ with respect to this volume. That is, we can form the integral

$$\int_M \sigma \cdot \tau \, \mathrm{vol}_x.$$

In this way, which depends on the metric x, we have associated to τ a continuous linear function ℓ_τ on $T_x^0 X$:

$$\ell_\tau(\sigma) = \int_M \sigma \cdot \tau \, \mathrm{vol}_x.$$

Here is a different kind of ℓ, one associated to a curve: Let I be a (compact) interval on the real line and $\mathbf{c} \colon I \to M$ a smooth non-degenerate curve on M. Let τ be a smooth tensor field along $\mathbf{c}$. This means that $\tau(s)$ is a contravariant 2-tensor at the point $\mathbf{c}(s)$. (We will assume that $\tau(s) \neq 0$ for any s.) If σ is a symmetric covariant 2-tensor, then

$$\sigma(\mathbf{c}(s)) \cdot \tau(s)$$

is a smooth function of s, and we can form the integral

$$\int_I \sigma(\mathbf{c}(s)) \cdot \tau(s)\,ds.$$

So, the pair consisting of the curve $\mathbf{c}$ and the tensor field τ along $\mathbf{c}$ gives rise to a continuous linear function $\ell_{c,\tau}$ on $T_x^0 X$ by

$$\ell_{c,\tau}(\sigma) := \int_I \sigma(\mathbf{c}(s)) \cdot \tau(s)\,ds.$$

We can now state the main result of Einstein, Infeld and Hoffman as reformulated by Souriau: If $\ell_{c,\tau}(\sigma)$ satisfies the condition

$$\ell_{c,\tau}(\sigma) \text{ vanishes on } T_x O$$

then up to a suitable reparametrization of $\mathbf{c}$, the curve $\mathbf{c}$ is a geodesic, and

$$\tau(s) = \pm \mathbf{c}'(s) \otimes \mathbf{c}'(s),$$

where $\mathbf{c}'$ denotes the tangent vector to $\mathbf{c}$.

The proof of this assertion can be found in my notes on Semi-Riemannian geometry and general relativity, which can be downloaded from my Harvard website (www.math.harvard.edu/people/SternbergShlomo. html). The argument consists essentially of repeated integrations by parts.

14. The Hilbert Function

For any metric x, let $S(x)$ denote the scalar curvature of x. Try to define the 'function' $F(x)$ by

$$F(x) = -\int_M S(x)\,\mathrm{vol}_x.$$

The trouble is that this integral may not be defined, since M, in general, is not compact.

Nevertheless, the variation

$$dF_x(\sigma)$$

for $\sigma \in T_x^0 X$ is well defined, since we may replace integration over M by integration over any compact subset K containing the support of σ and then define

$$dF_x(\sigma) = \frac{d}{dt}\int_K S(x + t\sigma)\,\mathrm{vol}_{x+t\sigma}\Big|_{|t=0}$$

This clearly does not depend on the choice of K.

Suppose ℓ is a linear function on $S_2^0(M)$ corresponding to a smooth tensor field in some (and hence every) metric. The Einstein–Hilbert field equations for the metric x are the equations

$$dF_x = \ell.$$

We know that a necessary condition for the solvability of these equations is

$$\ell \text{ vanishes on } T_x O.$$

15. The Passive Equations and the Einstein–Infeld– Hoffman–Souriau Solution to the 'Fundamentally Simple Question'

Suppose we replace ℓ by $\ell + \ell'$, where (say) ℓ' is a smooth approximation to $\ell_{\mathbf{c},\mu}$. Then we get, in principle, a different x' as a solution to

$$dF_{x'} = \ell + \ell'$$

and hence also a different orbit, O'. The 'passivity approximation' that I stated in Section 1 says that we will ignore this change in x and hence assume the necessary condition

$$\ell + \ell' \text{ vanishes on } T_x O.$$

Since this condition is linear, and we know that

$$\ell \text{ vanishes on } T_x O$$

we conclude that

$$\ell' \text{ vanishes on } T_x O$$

and hence (in the limit) that

$$\ell_{\mathbf{c},\mu} \text{ vanishes on } T_x O,$$

and so $\mathbf{c}$ is a geodesic. This is the Einstein–Infeld–Hoffman solution to their 'fundamentally simple question' as reformulated by Souriau.

16. The Schrödinger Equation

I will illustrate the 'integration by parts' argument that I omitted, by studying an analogue of this procedure in a finite dimensional model of 'quantum mechanics.' Let V be a finite dimensional vector space, and let $G = Gl(V)$

be the group of all invertible linear transformations of V. Let X be the space of all linear transformations of V with G acting on X by conjugation:

$$g \cdot x := gxg^{-1}.$$

The tangent space to the orbit O through x consists of all $[x, y]$ as y varies over X. Since X is a vector space, we can identify $T_x X$ with X for every x. We can also identify the space of linear functions on X with X using the trace: If $z \in X$ define

$$\ell_z(w) = \operatorname{tr} zw.$$

Every linear function on X is of this form. The condition

$$\ell_z \text{ vanishes on } T_x O$$

becomes

$$\operatorname{tr} z[x, y] = 0 \,\forall y \in X.$$

But

$$\operatorname{tr} zyx = \operatorname{tr} xzy,$$

so

$$\operatorname{tr} z[x, y] = \operatorname{tr} z(xy - yx) = \operatorname{tr}(zx - xz)y = \operatorname{tr} [z, x]y.$$

(This was the 'integration by parts.')

The condition $\operatorname{tr}[z, x]y = 0$ for all y implies that $[x, z] = 0$. So the condition

$$\ell_z \text{ vanishes on } T_x O$$

is

$$[x, z] = 0$$

in the current example.

Suppose we look at a special kind of z (like we considered the special ℓ associated to a curve in the general relativity case). Suppose z is of rank 1, and so maps the entire space V onto the line through a vector φ. Then the above condition implies that φ is an eigenvector of x:

$$x\varphi = \lambda\varphi$$

for some λ.

To write this in more familiar notation, replace the letter x with the letter H. We obtain

$$H\varphi = \lambda\varphi.$$

So the condition

$$\ell_z \text{ vanishes on } T_x O,$$

together with the assumption that we are looking at a special kind of ℓ, one corresponding to a rank 1 operator, gives Schrödinger's equation. We have derived Schrödinger's equation from the same principle that gave us the equation for geodesics!

Prof. Shlomo Sternberg delivered the Albert Einstein Memorial Lecture in 2006.

The Structure of Quarks and Leptons

Haim Harari

In the past 100 years the search for the ultimate constituents of matter has penetrated four layers of structure. All matter has been shown to consist of atoms. The atom itself has been found to have a dense nucleus surrounded by a cloud of electrons. The nucleus in turn has been broken down into its component protons and neutrons. More recently it has become apparent that the proton and the neutron are also composite particles; they are made up of the smaller entities called quarks. What comes next? It is entirely possible that the progression of orbs within orbs has at last reached an end and that quarks cannot be more finely divided. The leptons, the class of particles that includes the electron, could also be elementary and indivisible. Some physicists, however, are not at all sure the innermost kernel of matter has been exposed. They have begun to wonder whether the quarks and leptons too might not have some internal composition.

The main impetus for considering still another layer of structure is the conviction (or perhaps prejudice) that there should be only a few fundamental building blocks of matter. Economy of means has long been a guiding principle of physics, and it has served well up to now. The list of the basic constituents of matter first grew implausibly long toward the end of the 19th century, when the number of chemical elements, and hence the number of species of atoms, was approaching 100. The resolution of atomic structure solved the problem, and in about 1935 the number of elementary particles stood at four: the proton, the neutron, the electron and the neutrino. This parsimonious view of the world was spoiled in the 1950s and 1960s; it turned out that the proton and the neutron are representatives of a very large family of particles, the family now called hadrons. By the mid-1960s the number of fundamental forms of matter was again roughly 100. This time it was the quark model that brought relief. In the initial formulation of the model all hadrons could be explained as combinations of just three kinds of quarks.

*This article is reprinted from the *Scientific American*, April 1983, pp. 48–60.

Now it is the quarks and leptons themselves whose proliferation is beginning to stir interest in the possibility of a simpler scheme. Whereas the original model had three quarks, there are now thought to be at least 18, as well as six leptons and a dozen other particles that act as carriers of forces. Three dozen basic units of matter are too many for the taste of some physicists, and there is no assurance that more quarks and leptons will not be discovered. Postulating a still deeper level of organization is perhaps the most straightforward way to reduce the roster. All the quarks and leptons would then be composite objects, just as atoms and hadrons are, and would owe their variety to the number of ways a few smaller constituents can be brought together. The currently observed diversity of nature would be not intrinsic but combinatorial.

It should be emphasized that as yet there is no evidence quarks and leptons have an internal structure of any kind. In the case of the leptons, experiments have probed to within 10^{-16} centimeter and found nothing to contradict the assumption that leptons are pointlike and structureless. As for the quarks, it has not been possible to examine a quark in isolation, much less to discern any possible internal features. Even as a strictly theoretical conception, the subparticle idea has run into difficulty: No one has been able to devise a consistent description of how the subparticles might move inside a quark or a lepton and how they might interact with one another. They would have to be almost unimaginably small: If an atom were magnified to the size of the earth, its innermost constituents could be no larger than a grapefruit. Nevertheless, models of quark and lepton substructure make a powerful appeal to the aesthetic sense and to the imagination: They suggest a way of building a complex world out of a few simple parts.

Any theory of the elementary particles of matter must also take into account the forces that act between them and the laws of nature that govern the forces. Little would be gained in simplifying the spectrum of particles if the number of forces and laws were thereby increased. As it happens, there has been a subtle interplay between the list of particles and the list of forces throughout the history of physics.

In about 1800 four forces were thought to be fundamental: gravitation, electricity, magnetism and the short-range force between molecules that is responsible for the cohesion of matter. A series of remarkable experimental and theoretical discoveries then led to the recognition that electricity and magnetism are actually two manifestations of the same basic force, which was soon given the name electromagnetism. The discovery of atomic structure brought a further revision. Although an atom is electrically neutral

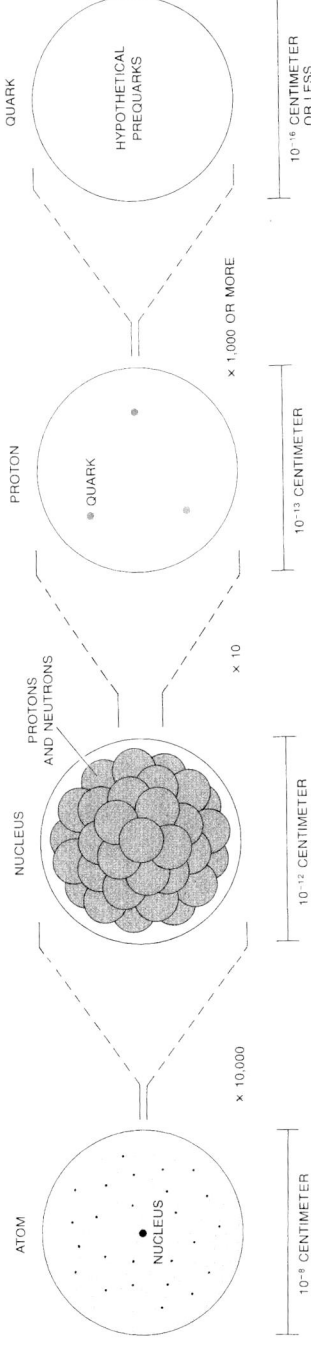

Figure 1. Hierarchy of particles in the structure of matter currently has four levels. All matter is made up of atoms; the atom consists of a nucleus surrounded by electrons; the nucleus is composed of protons and neutrons; each proton and neutron is thought to be composed of three quarks. Recent speculations might add a fifth level: The quark might be a composite of hypothetical finer constituents, which can be generically called prequarks. The leptons, the class of particles that includes the electron, could also consist of prequarks.

overall, its constituents are charged, and the short-range molecular force came to be understood as a complicated residual effect of electromagnetic interactions of positive nuclei and negative electrons. When two neutral atoms are far apart, there are practically no electromagnetic forces between them. When they are near each other, however, the charged constituents of one atom are able to 'see' and influence the inner charges of the other, leading to various short-range attractions and repulsions.

As a result of these developments physics was left with only two basic forces. The unification of electricity and magnetism had reduced the number by one, and the molecular interaction had been demoted from the rank of a fundamental force to that of a derivative one. The two remaining fundamental forces, gravitation and electromagnetism, were both long-range. The exploration of nuclear structure, however, soon introduced two new short-range forces. The strong force binds protons and neutrons together in the nucleus, and the weak force mediates certain transformations of one particle into another, as in the beta decay of a radioactive nucleus. Thus there were again four forces.

The development of the quark model and the accompanying theory of quark interactions was the next occasion for revising the list of forces. The quarks in a proton or a neutron are thought to be held together by a new long-range fundamental force called the color force, which acts on the quarks because they bear a new kind of charge called color. (Neither the force nor the charge has any relation to ordinary colors.) Just as an atom is made up of electrically charged constituents but is itself neutral, so a proton or a neutron is made up of colored quarks but is itself colorless. When two colorless protons are far apart, there are essentially no color forces between them, but when they are near, the colored quarks in one proton 'see' the color charges in the other proton. The short-range attractions and repulsions that result have been identified with the effects of the strong force. In other words, just as the short-range molecular force became a residue of the long-range electromagnetic force, so the short-range strong force has become a residue of the long-range color force.

One more chapter can be added to this abbreviated history of the forces of nature. A deep and beautiful connection has been found between electromagnetism and the weak force, bringing them almost to the point of full unification. They are clearly related, but the connection is not quite as close as it is in the case of electricity and magnetism, and so they must still be counted as separate forces. Therefore the current list of fundamental forces still has four entries: the long-range gravitational, electromagnetic and color

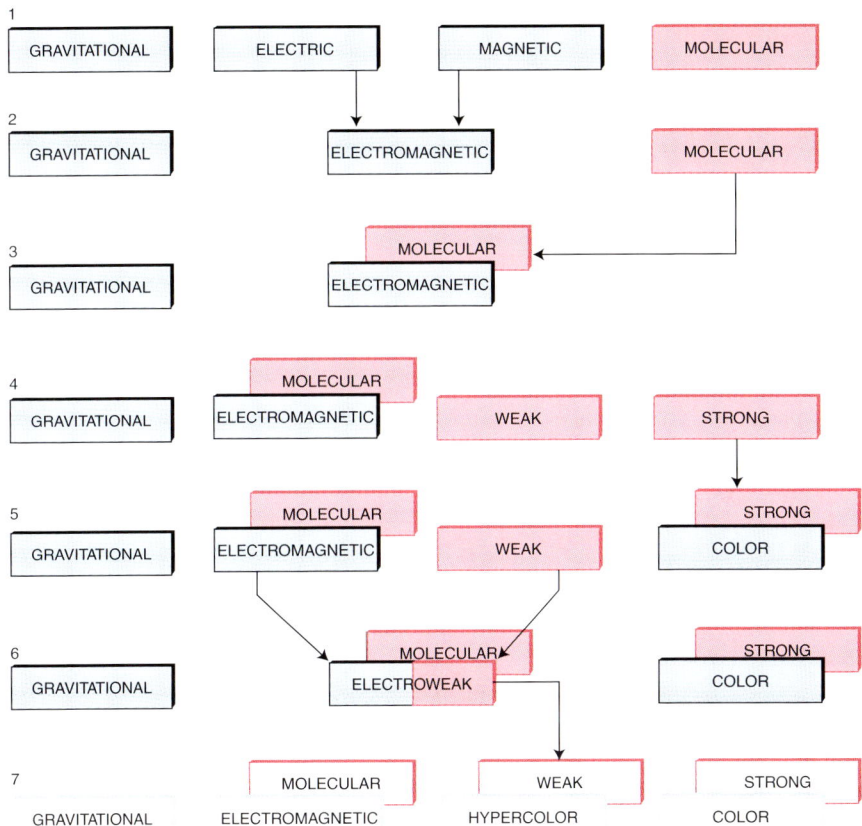

Figure 2. FUNDAMENTAL FORCES OF NATURE can be classified in a scheme that has evolved together with the list of elementary particles. Here long-range forces are shown in gray and short-range ones in color. Early in the 19th century three long-range forces were thought to be fundamental: the gravitational, the electric and the magnetic. One short-range force, the molecular force responsible for the cohesion of matter, also had fundamental status. James Clerk Maxwell unified electricity and magnetism, and with the discovery of atomic structure it became apparent that the molecular force is not fundamental but instead is a residual effect of electromagnetism. The discovery of the atomic nucleus introduced two short-range forces, the weak and the strong. In the quark model, however, the strong force becomes a residue of a new long-range force called the color force. Furthermore, a deep relation has been found between the weak force and electromagnetism, so that they can be considered partially unified. The sixth row of the chart represents the forces of nature as they are now understood in the 'standard model' of elementary-particle physics. A successful model of quark and lepton structure could bring a further revision. In some models, for example, there is a new long-range force called hypercolor, and the weak force is a residue of it. In models of this kind all the fundamental forces of nature are long-range ones. Such models, however, are still highly speculative.

forces and the short-range weak force. Within the limits of present knowledge all natural phenomena can be understood through these forces and their residual effects.

The evolution of ideas about particles and that of ideas about forces are clearly interdependent. As new basic particles are found, old ones turn out to be composite objects. As new forces are discovered, old ones are unified or reduced to residual status. The lists of particles and forces are revised from time to time as matter is explored at smaller scale and as theoretical understanding progresses. Any change in one list inevitably leads to a modification of the other. The recent speculations about quark and lepton structure are no exception; they too call for changes in the complement of forces. Whether the changes represent a simplification remains to be seen.

Of the four established fundamental forces, gravitation must be put in a category apart. It is too feeble even to be detected in the interactions of individual particles, and it is not understood in terms of microscopic events. For the other three forces successful theories have been developed and are now widely accepted. The three theories are distinct, but they are consistent with one another; taken together they constitute a comprehensive model of elementary particles and their interactions, which I shall refer to as the standard model.

In the standard model the indivisible constituents of matter are the quarks and the leptons. It is convenient to discuss the leptons first. There are six of them: the electron and its companion the electron-type neutrino, the muon and the muon-type neutrino, and the tau and the tau-type neutrino. The electron, the muon and the tau have an electric charge of -1; the three neutrinos are electrically neutral.

There are also six basic kinds of quark, which have been given the names up, down, charmed, strange, top and bottom, or u, d, c, s, t and b. (The top quark has not yet been detected experimentally, and neither has the tau-type neutrino, but few theorists doubt their existence.) The u, c and t quarks have an electric charge of $+2/3$, the d, s and b quarks a charge of $-1/3$. In addition each quark type has three possible colors, which I shall designate red, yellow and blue. Thus if each colored quark is counted as a separate particle, there are 18 quark varieties altogether. Note that each quark carries both color and electric charge, but none of the leptons are colored.

For each particle in this scheme there is an antiparticle with the same mass but with opposite values of electric charge and color. The antiparticle of the electron is the positron, which has a charge of $+1$. The antiparticle of a red u quark, with a charge of $+2/3$, is an antired $\bar{u}$ antiquark, with a charge of $-2/3$.

The color property of the quarks is analogous in many ways to electric charge, but because there are three possible colors it is appreciably more complicated. Electrically charged particles can be brought together to form an electrically neutral system in only one way: by combining equal quantities of positive and negative charge. A colorless composite particle can be

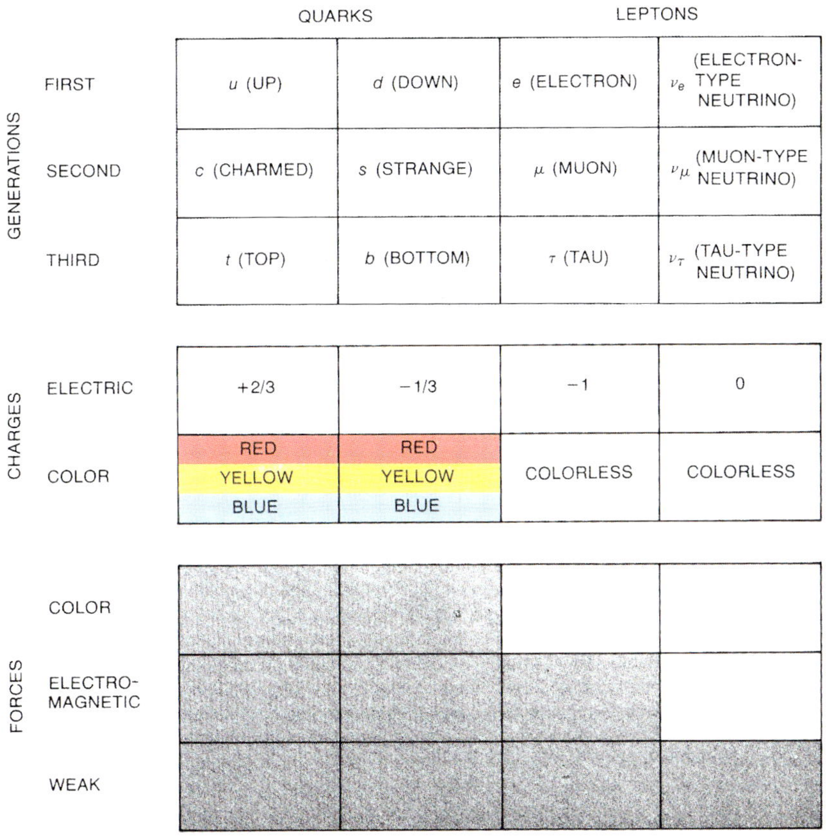

		QUARKS		LEPTONS	
GENERATIONS	FIRST	u (UP)	d (DOWN)	e (ELECTRON)	ν_e (ELECTRON-TYPE NEUTRINO)
	SECOND	c (CHARMED)	s (STRANGE)	μ (MUON)	ν_μ (MUON-TYPE NEUTRINO)
	THIRD	t (TOP)	b (BOTTOM)	τ (TAU)	ν_τ (TAU-TYPE NEUTRINO)

CHARGES	ELECTRIC	$+2/3$	$-1/3$	-1	0
	COLOR	RED YELLOW BLUE	RED YELLOW BLUE	COLORLESS	COLORLESS

FORCES	COLOR				
	ELECTRO-MAGNETIC				
	WEAK				

Figure 3. Standard model of elementary particles includes three 'generations' of quarks and leptons, although all ordinary matter can be constructed out of the particles of the first generation alone. The quarks are distinguished by fractional values of electric charge and by a property that is fancifully called color: Each quark type comes in red, yellow and blue versions. The leptons have integer units of electric charge and are colorless. The two classes of particles also differ in their response to the various forces. Only the quarks are subject to the color force, and as a result they may be permanently confined inside composite particles such as the proton.

formed out of colored quarks in much the same way, namely by combining a colored quark and an anticolored antiquark. In the case of color, however, there is a second way to form a neutral state: any composite system with equal quantities of all three colors or of all three anti-colors is also colorless. For this reason a proton consisting of one red quark, one yellow quark and one blue quark has no net color.

One further property of the quarks and leptons should be mentioned: each particle has a spin, or intrinsic angular momentum, equal to one-half the basic quantum-mechanical unit of angular momentum. When a particle with a spin of 1/2 moves along a straight line, its intrinsic rotation can be either clockwise or counterclockwise when the particle is viewed along the direction of motion. If the spin is clockwise, the particle is said to be right-handed, because when the fingers of the right hand curl in the same direction as the spin, the thumb indicates the direction of motion. For a particle with the opposite sense of spin a left-hand rule describes the motion, and so the particle is said to be left-handed.

In the standard model the three forces that act on the quarks and leptons are described by essentially the same mathematical structure. It is known as a gauge-invariant field theory or simply a gauge theory. Each force is transmitted from one particle to another by carrier fields, which in turn are embodied in carrier particles, or gauge bosons.

The gauge theory of the electromagnetic force, called quantum electrodynamics or QED, is the earliest and simplest of the three theories. It was devised in the 1940s by Richard P. Feynman, Julian S. Schwinger and Sin-Itiro Tomonaga. QED describes the interactions of electrically charged particles, most notably the electron and the positron. There is one kind of gauge boson to mediate the interactions; it is the photon, the familiar quantum of electromagnetic radiation, and it is massless and has no electric charge of its own. QED is probably the most accurately tested theory in physics. For example, it correctly predicts the magnetic moment of the electron to at least 10 significant digits.

The theory of the color force was formulated by analogy to QED and is called quantum chromodynamics or QCD. It was developed over a period of almost two decades through the efforts of many theoretical physicists. In QCD particles interact by virtue of their color rather than their electric charge. The gauge bosons of QCD, which are responsible for binding quarks inside a hadron, are called gluons. Like the photon, the gluons are massless, but whereas there is just one kind of photon, there are eight species of gluons.

A further difference between the photon and the gluons turns out to be even more important. Although the photon is the intermediary of the electromagnetic force, it has no electric charge and hence gives rise to no electromagnetic forces of its own (or at least none of significant magnitude). The gluons, in contrast, are not colorless. They transmit the color force between quarks but they also have color of their own and respond to the color force. This reflexiveness, whereby the carrier of the force acts on itself, makes a complete mathematical analysis of the color force exceedingly difficult.

One peculiarity that seems to be inherent in QCD is the phenomenon of color confinement. It is thought that the color force somehow traps colored objects (such as quarks and gluons) inside composite objects that are invariably colorless (such as protons and neutrons). The colored particles can never escape (although they can form new colorless combinations). It is because of color confinement, physicists suppose, that a quark or a gluon has never been seen in isolation. I must stress that although the idea of color confinement is now widely accepted, it has not been proved to follow from QCD. There may still be surprises in store.

The weak force is somewhat different from the other two, but it can nonetheless be described by a gauge theory of the same general kind. The theory was worked out, and the important connection between the weak force and electromagnetism was established, in the 1960s and the early 1970s by a large number of investigators. Notable contributions were made (in chronological order) by Sheldon Lee Glashow of Harvard University, Steven Weinberg of the University of Texas at Austin, Abdus Salam of the International Centre for Theoretical Physics in Trieste and Gerard 't Hooft of the University of Utrecht.

Curiously, the charges on which the weak force acts are associated with the handedness of a particle. Among both quarks and leptons left-handed particles and right-handed antiparticles have a weak charge, but right-handed particles and left-handed antiparticles are neutral with respect to the weak force. What is odder still, the weak charge is not conserved in nature: a unit of charge can be created out of nothing or can disappear into the vacuum. In contrast, the net quantity of electric charge in an isolated system of particles can never be altered, and neither can the net color. The weak force is also distinguished by its exceedingly short range; its effects extend only to a distance of about 10^{-16} centimeter, or roughly a thousandth of the diameter of a proton.

In the gauge theory of the weak force both the failure of the weak charge to be conserved and the short range of the force are attributed to a mechanism

called spontaneous symmetry breaking, which I shall discuss in greater
detail below. For now it is sufficient to note that the symmetry-breaking
mechanism implies that the weak charge, and the associated handedness of
particles, should be conserved at extremely high energy, where a particle's
mass is a negligible fraction of its kinetic energy.

Spontaneous symmetry breaking also requires that the gauge bosons of
the weak force be massive particles; indeed, they have masses approximately
100 times the mass of the proton. In the standard model there are three
such bosons: Two of them, designated W^+ and W^-, carry electric charge
as well as weak charge; the third, designated Z^0, is electrically neutral. The
large mass of the weak bosons accounts for the short range of the force.
According to the uncertainty principle of quantum mechanics, the range of
a force is inversely proportional to the mass of the particle that transmits
it. Thus electromagnetism and the color force, being carried by massless
gauge bosons, are effectively infinite in range, whereas the weak force has an
exceedingly small sphere of influence. Spontaneous symmetry breaking has
still another consequence: It predicts the existence of at least one additional
massive particle, separate from the weak bosons. It is called the Higgs
particle after Peter Higgs of the University of Edinburgh, who made an
important contribution to the theory of spontaneous symmetry breaking.

In the past 10 years the successes of the standard model have given physicists a good deal of self-confidence. All known forms of matter can be constructed out of the 18 colored quarks and the six leptons of the model. All
observed interactions of matter can be explained as exchanges of the 12 gauge
bosons included in the model: the photon, the eight gluons and the three weak
bosons. The model seems to be internally consistent; no one part is in conflict with any other part, and all measurable quantities are predicted to have
a plausible, finite value. Internal consistency is not a trivial achievement in
a conceptual system of such wide scope. So far the model is also consistent
with all experimental results, that is to say, no clear prediction of the model
has yet been contradicted by experiment. To be sure, there are some important predictions that have not yet been fully verified; most notably, the tau-
type neutrino, the top quark, the weak bosons and the Higgs particle must be
found. The first direct evidence of W bosons was recently reported by a group
of experimenters at CERN, the European Laboratory for Particle Physics in
Geneva. In the next several years new particle accelerators and more sensitive detecting apparatus will test the remaining predictions of the model. Most
physicists are quite certain they will be confirmed.

If the standard model has proved so successful, why would anyone consider more elaborate theories? The primary motivation is not a suspicion

that the standard model is wrong but rather a feeling that it is less than fully satisfying. Even if the model gives correct answers for all the questions it addresses, many questions are left unanswered and many regularities in nature remain coincidental or arbitrary. In short, the model itself stands in need of explanation.

The strongest hint of some organizing principle beyond the standard model is the proliferation of elementary particles. The known properties of matter are not so numerous or diverse that 24 particles are needed to represent them all. Indeed, there seems to be a great deal of repetition in the spectrum of quarks and leptons. There are three leptons with an electric charge of -1, three neutral leptons, three quarks with a charge of $+2/3$ and three quarks with a charge of $-1/3$. Everything is triplicated, and for no apparent reason. A world constructed by choosing one particle from each of the four groups would seem to have all the necessary variety.

As it turns out, all ordinary matter can indeed be formed from a subset that includes just the u quark, the d quark, the electron and the electron-type neutrino. These four particles and their antiparticles make up the 'first generation' of quarks and leptons. The remaining quarks and leptons merely repeat the same pattern in two additional generations without seeming to add anything new. Corresponding particles in different generations are identical in all respects except one: They have different masses. The d, s and b quarks, for example, respond in precisely the same way to the electromagnetic, color and weak forces. For some unknown reason, however, the s quark is roughly 20 times as heavy as the d quark, and the b quark is approximately 600 times as heavy as the d. The mass ratios of the other quarks and of the charged leptons are likewise large and unexplained. (The masses of the neutrinos are too small to have been measured; it is not yet known whether the neutrinos are merely very light or are entirely massless.)

The presence of three generations of quarks and leptons begs for an explanation. Why does nature repeat itself? The pattern of particle masses is also mysterious. In the standard model the masses are determined by approximately 20 'free' parameters that can be assigned any values the theorist chooses; in practice the values are generally based on experimental findings. Is it possible the 20 parameters are all unrelated? Are they fundamental constants of nature with the same status as the velocity of light or the electric charge of the electron? Probably not.

A further tantalizing regularity can be perceived in the electric charges of the quarks and leptons: They are all related by simple ratios and are all integer multiples of one-third the electron charge. The standard model

ELECTRIC CHARGE

	+1	+2/3	+1/3	0	−1/3	−2/3	−1
ANTILEPTONS	$\bar{e}$			$\bar{\nu}_e$			
QUARKS		u			d		
ANTIQUARKS			$\bar{d}$			$\bar{u}$	
LEPTONS				ν_e			e

Figure 4. First generation of quarks and leptons forms an orderly pattern when the particles are arranged according to their electric charge. All values of charge from +1 to −1 in intervals of 1/3 are represented. All colored particles have fractional charge and all colorless ones have integral charge. The pattern is an arbitrary feature of the standard model, where charge and color are independent, but it might have some explanation if quarks and leptons are composite.

supplies no reason; in principle the charge ratios could have any values. It can be deduced from observation that the ratios of one-third and two-thirds that define the quark charges are not approximations. The proton consists of two u quarks and a d quark, with charges of $2/3 + 2/3 - 1/3$, or +1. If these values were not exact and the quarks instead had charges of, say, +0.617 and −0.383, the magnitude of the proton's charge would not be exactly equal to that of the electron's, and ordinary atoms would not be electrically neutral. Since atoms can be brought together in enormous numbers, even a slight departure from neutrality could be readily detected.

If the particles and antiparticles that make up a single generation are arranged according to their charge, it is found that every value from −1 to +1 in intervals of one-third is occupied by one particle (or, in the case of zero charge, by two particles, namely the neutrino and the antineutrino). The pattern formed raises still more questions. Why has nature favored these values of electric charge but no others, such as +4/3 or −5/3? It is apparent that all particles with integral charge are colorless and all those with fractional charge are colored. Is there some relation between the electric charge of a particle and its color or between the quarks and the leptons? The standard model implies no such relations, but they seem to exist.

Another motivation for looking beyond the standard model is the continuing desire to unify the fundamental forces, or at least to find some relation among them. The cause of parsimony would be served, for example, if two of the forces could be consolidated, as electricity and magnetism were, or if one force could be made a residue of another, as the strong force was made a residue of the color force. Ironically, it may turn out that a simplification of this kind can be attained only by introducing still more forces.

A theory that 'goes beyond' the standard model need not contradict or invalidate it. The standard model may emerge as a very good approximation of the deeper theory. The standard model gives a remarkably successful description of all phenomena at distances no smaller than about 10^{-16} centimeter. A deeper theory should therefore focus on events at a still smaller scale. If there are new constituents to be discovered, they must exist within such minuscule regions of space. If there are new forces, their action must be effective only at a distance of less than 10^{-16} centimeter, either because the force is inherently short-range (following the example of the weak force) or because it is subject to some form of confinement (as the color force is).

The search for a theory beyond the standard model was launched almost 10 years ago, and by now several directions have been explored. One promising direction has led to the models known as grand unified theories, which incorporate the electromagnetic, color and weak forces into one fundamental force. The essential idea is to put all the quarks and leptons that make up one generation into a single family; new gauge bosons are then postulated to mediate interactions between the colored quarks and the colorless leptons. The theories account for the regularities noted in the distribution of electric charge and explain the exact commensurability of the quark and lepton charges. On the other hand, they do nothing to reduce the number of fundamental constants, they shed no light on the triplication of the generations and they create certain new theoretical difficulties of their own.

There have been several variations on the theme of grand unification. For example, the concept of horizontal symmetry tackles the triplication problem by establishing a symmetry relation among the generations. The mathematically beautiful idea called supersymmetry relates particles whose spin angular momentum is a half-integer (such as the quarks and leptons) to those with integer spin (such as the gauge bosons). The technicolor theory suggests that the Higgs particle of the standard model is a composite object made up of new fundamental entities; they would be bound together by a new force analogous to the color force and called technicolor. Each of these ideas answers some of the questions that remain open in the standard model. Each idea also fails to answer other questions, raises new difficulties and worsens existing ones, for example by further increasing the number of unrelated arbitrary constants.

In all the above schemes for grand unification it is explicitly assumed that the quarks, the leptons, the photon, the gluons and the weak bosons are the truly fundamental particles of the ultimate theory of nature. The alternative of suggesting that the quarks and leptons are themselves composite is in one sense the most conservative and the least original hypothesis. It is a strategy that has worked before, repeatedly, in going from the atom to the nucleus to the proton to the quark. In another sense the idea of quark and lepton substructure is a most radical proposal. The electron has now been studied for almost a century, and its pointlike nature has been established very well indeed. In the case of the neutrino, which may turn out to be entirely massless, it is even more difficult to imagine an internal structure. The assertion that these particles and the others like them are composites will clearly have to overcome formidable obstacles if it is to have any future.

Offsetting the difficulties of the undertaking are its potential rewards. A fully successful composite model might resolve all the questions left unsettled in the standard model. Such a hypothetical theory would begin by introducing a new set of elementary particles, which I shall refer to generically as prequarks. Ideally there would not be too many of them. Each quark and lepton in the standard model would be accounted for as a combination of prequarks, just as each hadron can be explained as a combination of quarks. The mass of a quark or a lepton would no longer be an arbitrary constant of nature; instead it would be determined by the masses of the constituent prequarks and by the strength of the force that binds the prequarks together. The exact ratios that relate the charge of a quark to that of a lepton would be explained in a similar way: Both kinds of composite particles would derive their charges from those of the same constituent prequarks. The entire pattern of quarks and leptons within a generation would presumably reflect some simple rules for combining the prequarks.

The existence of multiple generations might also be explained in a natural way. The quarks and leptons in the higher generations might have an internal constitution similar to that of the corresponding particles of the first generation; the differences could be in the energy and the state of motion of the constituents. Thus the s and b quarks would be excited states of the d quark, and the muon and the tau lepton would be excited states of the electron. Similar excited states are known in all other composite systems, including atoms, nuclei and hadrons. For example, at least a dozen hadrons have been identified in experiments as excited states of the proton; they and the proton itself are all thought to have essentially the same quark composition, namely *uud*.

This imaginary, ideal prequarks theory accomplishes everything one might ask of it except for unifying the fundamental forces. Even there some progress is conceivable, since a new force would very likely be introduced to bind the prequarks together; the new force might lead to a new understanding of how the known forces are related. Imagining what a successful model might be like, however, is not at all the same thing as actually constructing a realistic and internally consistent one. So far no one has done it.

What has been lacking is a satisfactory theory of prequark dynamics, a theory that would describe how the prequarks move inside a quark or a lepton and that would enable one to calculate the mass and total energy of the system. As I shall set forth below, there are fundamental obstacles to the formulation of such a theory, although I would submit that they are not insurmountable. In the meantime, lacking any persuasive account of prequark motions, theorists have nonetheless been exploring the combinatorial possibilities of the prequark idea, that is, they have been examining the ways quarks and leptons might be built up as specific combinations of finer constituents.

In the past few years several dozen composite models have been proposed; they can be classified in perhaps four or five main groups. No single model solves all problems, answers all questions and is widely accepted. It would be unfair to describe only one scheme, but it is impractical to enumerate them all. I shall present a few of the central ideas.

The first explicit model of quark and lepton substructure was proposed in 1974 by Jogesh C. Pati of the University of Maryland at College Park and Salam, who have since returned to the topic several times in collaboration with John Strathdee of the International Centre for Theoretical Physics. It was they who introduced the term prequark, which I have adopted here as a generic name for hypothetical subconstituents of all kinds. The specific elementary particles of the model devised by Pati and Salam I shall call preons, which is another term of their invention.

The rationale for the preon model begins with the observation that every quark and lepton can be identified unambiguously by listing just three of its properties: electric charge, color and generation number. These properties, then, suggest a straightforward way of organizing a set of constituent particles. Three families of preons are needed. In one family the preons carry electric charge, in another they carry color and in the third they have some property that determines generation number. A given quark or lepton is assembled by selecting exactly one preon from each family.

The preons that determine generation number are called somons, from the Greek *soma*, meaning body, because they have a dominant influence on the mass of the composite system. Since there are three generations of quarks and leptons, there must be three somons. The color of the composite system is determined by preons called chromons; there are four of them, one with the color red, one yellow, one blue and one colorless. The remaining family of preons, which is assigned the role of defining electric charge, needs to have only two members in order for every quark and lepton to be uniquely identified. These last preons have been given the name flavons, after flavor, the whimsical term for whatever property it is that distinguishes the u quark from the d quark, the c from the s, the neutrino from the electron and so on.

In the preon model the classification of a composite particle follows directly from its complement of preons. All leptons, for example, are distinguished by a colorless chromon, and all first-generation particles must obviously have a first-generation somon. In the allocation of electric charge, however, a complication arises. If there are only two flavons and if they are the sole carriers of electric charge, not all the charge values observed in nature can be reproduced. The u quark and the neutrino, for example, must have the same charge (because they include the same flavon), and so must the d quark and the electron. The problem can be solved in any of several ways. In one scheme electric charge is assigned to both the flavons and the chromons, and the total charge of a composite particle is equal to the sum of the two values. Models of this kind can be made to yield the correct charge states, but only by abandoning the principle of having each kind of preon carry just one property.

	PREON	ELECTRIC CHARGE	COLOR	GENERATION NUMBER
FLAVONS	f_1	+1/2	COLORLESS	0
	f_2	−1/2	COLORLESS	0
CHROMONS	c_R	+1/6	RED	0
	c_Y	+1/6	YELLOW	0
	c_B	+1/6	BLUE	0
	c_0	−1/2	COLORLESS	0
SOMONS	s_1	0	COLORLESS	1
	s_2	0	COLORLESS	2
	s_3	0	COLORLESS	3

Figure 5. Preon model assigns three properties of quarks and leptons to three groups of hypothetical constituents called flavons, chromons and somons. A quark or a lepton is formed by choosing one preon from each group. The flavons have the primary responsibility for determining electric charge, the chromons determine color and the somons determine generation number. Ideally each kind of preon would carry just one property, but some adjustment is needed to differentiate the fractional electric charges of the quarks from the integral charges of the leptons. In the version of the model shown here the chromons carry electric charge as well as color.

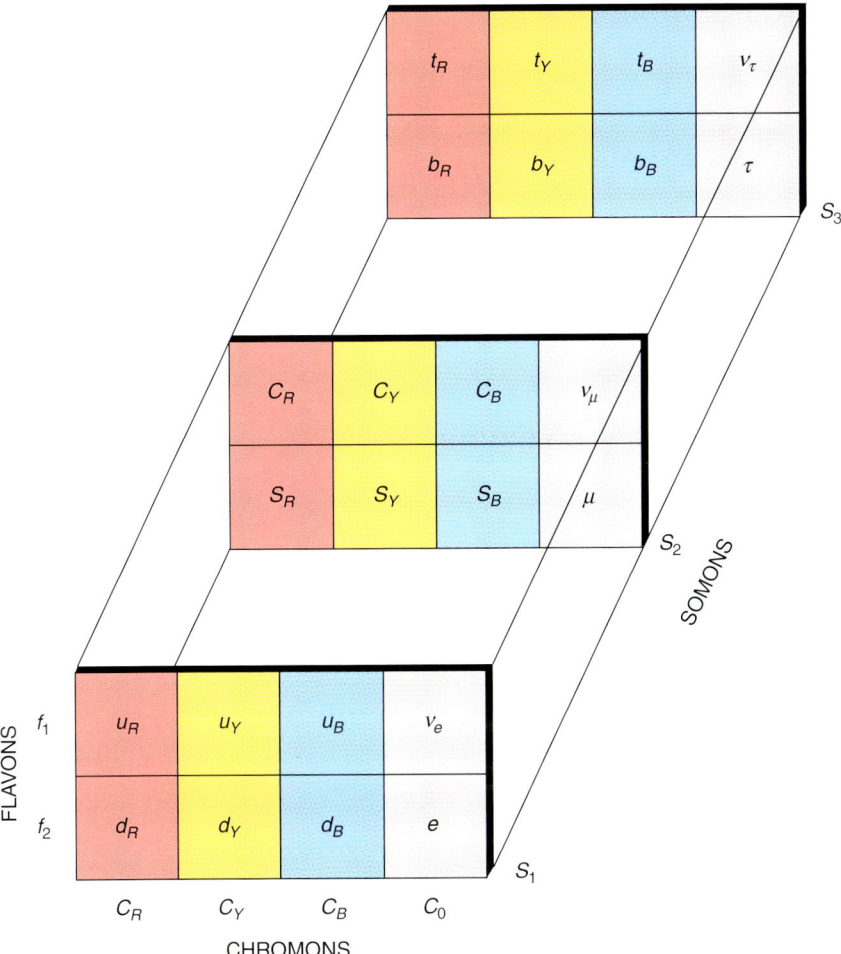

Figure 6. Combinations of preons give rise to the 24 quarks and leptons of the three generations. For example, the red s quark is made up of somon S_2 (signifying that the composite is second-generation particle), in combination with the red chromon and the negative flavon.

Another troublesome feature of the preon model is the requirement that composites be formed only by drawing one preon from each family. Why are there no particles made up of three chromons, say, or of two somons and a flavon? The exotic properties of such particles would make them quite conspicuous. It seems likely that if they existed, they would have been detected by now.

Many variations of the preon model have been proposed by other physicists, using the same basic idea but slightly different sets of preons. Notable among the variations are the models suggested by Hidezumi Terazawa, Yoichi Chikashige and Keiichi Akama of the University of Tokyo and by O. Wallace Greenberg and Joseph Sucher of the University of Maryland.

Perhaps the simplest model of quark and lepton structure is the rishon model, which I proposed in 1979. A similar idea was put forward at about the same time by Michael A. Shupe of the University of Illinois at Urbana-Champaign. The model has since been further developed and studied in great detail by Nathan Seiberg and me at the Weizmann Institute of Science in Rehovot. The model postulates just two species of fundamental building blocks, called rishons. (*Rishon* is the Hebrew adjective meaning first or primary.) One rishon has an electric charge of $+1/3$ and the other is electrically neutral. I designate them respectively T and V, for *Tohu Vavohu*, Hebrew for 'formless and void,' the description of the initial state of the universe given in the first chapter of Genesis. The complementary antirishons have charges of $-1/3$ and zero and are designated $\bar{T}$ and $\bar{V}$.

The model has one simple rule for constructing a quark or a lepton: Any three rishons can be assembled to form a composite system, or any

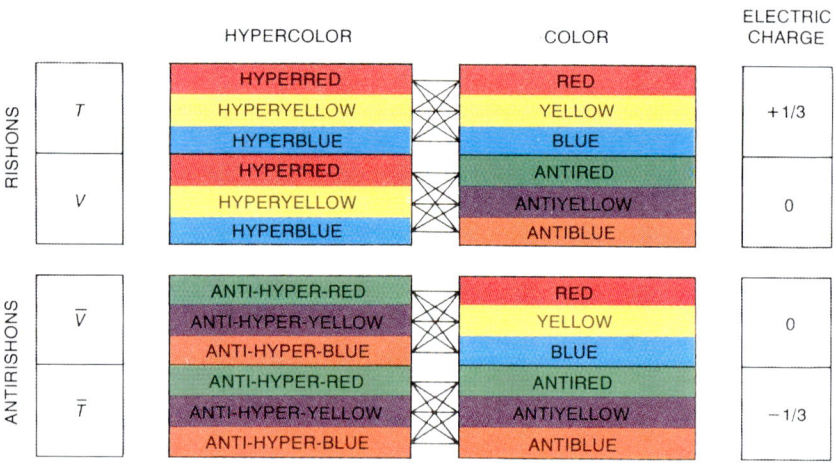

Figure 7. Rishon model constructs all the quarks and leptons out of just two species of fundamental particles and their antiparticles. The rishons carry both hypercolor, a property associated with the force that binds them to one another, and ordinary color, which they convey to the composite systems they form. One rishon is electrically charged and the other is neutral.

three antirishons, but rishons and antirishons cannot be mixed in a single particle. The rule gives rise to 16 combinations, which reproduce exactly the properties of the 16 quarks, antiquarks, leptons and antileptons in the first generation. In other words, every quark and lepton in the first generation corresponds to some allowed combination of rishons or antirishons. (In this system of classification each color is counted separately.)

The pattern of quark and lepton charges is generated as follows. The TTT combination, with rishon charges of $1/3 + 1/3 + 1/3$, has a total charge of $+1$ and therefore corresponds to the positron; similarly, $\overline{T}\overline{T}\overline{T}$ has

RISHON COMBINATION	PARTICLE	COLOR	ELECTRIC CHARGE
TTT	$\bar{e}$	COLORLESS	$+1$
TTV	u	RED YELLOW BLUE	$+2/3$
TVV	$\bar{d}$	ANTIRED ANTIYELLOW ANTIBLUE	$+1/3$
VVV	ν_e	COLORLESS	0

$\overline{V}\overline{V}\overline{V}$	$\bar{\nu}_e$	COLORLESS	0
$\overline{V}\overline{V}\overline{T}$	d	RED YELLOW BLUE	$-1/3$
$\overline{V}\overline{T}\overline{T}$	$\bar{u}$	ANTIRED ANTIYELLOW ANTIBLUE	$-2/3$
$\overline{T}\overline{T}\overline{T}$	e	COLORLESS	-1

Figure 8. Combinations of rishons taken three at a time give a correct accounting of all the quarks and leptons (and antiquarks and antileptons) in any one generation. The pattern of electric charges noted in the standard model, and the apparent relation between fractional charge and color, emerge as natural consequences of the way the rishons combine. All the allowed combinations of three rishons or of three antirishons are neutral with respect to hypercolor.

a total charge of -1 and is identified with the electron. The VVV and $\bar{V}\bar{V}\bar{V}$ combinations are both electrically neutral and represent the neutrino and the antineutrino respectively. The remaining allowed combinations yield fractionally charged quarks. TTV, with a charge of $+2/3$, is the u quark, and TVV, with a charge of $+1/3$, is the $\bar{d}$ antiquark. The analogous antirishon states $\bar{V}\bar{V}\bar{T}$ and $\bar{V}\bar{T}\bar{T}$ correspond to the d quark and the $\bar{u}$ antiquark.

The model also accounts successfully for the color of the composite systems. A T rishon can have any of the three colors red, yellow and blue, whereas a V rishon has an anticolor. Combinations such as $\bar{T}\bar{T}\bar{T}$ and VVV, which designate leptons, can be made colorless since they can include one rishon in each color or one in each anticolor. The other combinations, which yield quarks, must have a net color. For example, a TTV state might have the rishon colors red, blue and antiblue; the antiblue would cancel the blue, leaving the system with a net color of red. In this way the connection between color and electric charge, which was apparent but unexplained in the standard model, is readily understood. Because of the way electric charge and color are allotted to the rishons, all composite systems with fractional charge turn out to be colored, and all systems with an integer charge can be made colorless.

Other regularities of the standard model also lose their air of mystery when rishons are introduced. Consider the hydrogen atom, made up of a proton and an electron, or in terms of quarks and leptons two u quarks, a d quark and an electron. The total rishon content of the quarks is four T's, one $\bar{T}$, two V's and two $\bar{V}$'s. The electric charge of the $\bar{T}$ cancels the charge of one T rishon, and the V's and $\bar{V}$'s also cancel (they have no charge in any case), leaving the proton with a net charge equal to that of a TTT system. The electron's rishon content is just the opposite: $\bar{T}\bar{T}\bar{T}$. Thus it is evident why the proton and the electron have charges of equal magnitude and why the hydrogen atom is neutral: The ultimate sources of the charge are pairs of matched particles and antiparticles.

The rishon model and many other models that explain the pattern of the first generation have difficulty accounting for the second and third generations. It would seem that such models lend themselves well to the scheme of forming each particle in the higher generations as an excited state of the corresponding particle in the first generation. The simplest idea would be to describe the muon, for example, as having the same prequark constituents as the electron, but in the muon the prequarks would have some higher-energy configuration. It is an elegant idea but, regrettably, it appears to be unworkable. The scheme implies differences in energy between the successive excited states that are much larger than the actual differences. The flaw is a fundamental one, and there seems to be no remedy.

Other possible mechanisms for creating multiple generations have been considered. Several physicists have suggested that the higher-generation relatives of a given state might be created by adding a Higgs particle, the 'extra' particle associated with the weak bosons in the standard model. Because a Higgs particle has no electric charge or color or even spin angular momentum, adding one to a composite system would alter only the mass. Hence an electron might be converted into a muon by adding one Higgs particle or into a tau by adding two or more Higgs particles. Seiberg and I have proposed another possible mechanism: A higher-generation particle could be formed by the addition of pairs of prequarks and antiprequarks. All charges and other properties must cancel in such a pair, and so again only the mass would be affected.

These ideas are currently at the stage of unrestrained speculation. No one knows what distinguishes the three generations from one another, or why there are three or whether there may be more. No explanation can be given of the mass difference between the generations. In short, the triplication of the generations is still a major unsolved puzzle.

A third kind of substructure model deserves mention. It tries to relate the possibility of quark and lepton structure to another fundamental problem: understanding the relativistic quantum theory of gravitation. Ideas of this kind have been explored by John Ellis, Mary K. Gaillard, Luciano Maiani and Bruno Zumino of CERN. One approach to their ideas is to consider the distances at which prequarks interact: The experimental limit is less than 10^{-16} centimeter, but the actual distance could be several orders of magnitude smaller still. At about 10^{-34} centimeter the gravitational force becomes strong enough to have a significant effect on individual particles. If the scale of the prequark interactions is this small, gravitation cannot be neglected. Ellis, Gaillard, Maiani and Zumino have outlined an ambitious program that aims to unify all the forces, including gravitation, in a scheme that treats not only the quarks and leptons but also the gauge bosons as composite particles. Like other composite models, however, this one has serious flaws.

Any prequark model, regardless of its details, must supply some mechanism for binding the prequarks together. There must be a powerful attractive force between them. One strategy is to postulate a new fundamental force of nature analogous in its workings to the color force of the standard model.

To emphasize the analogy the new force is called the hypercolor force and the carrier fields are called hypergluons. The prequarks are assumed to have hypercolor, but they combine to form hypercolorless composite systems, just as quarks have ordinary color but combine to form colorless protons and neutrons. The hypercolor force presumably also gives rise to the property of confinement, again in analogy to the color force. Hence all hypercolored prequarks would be trapped inside composite particles, which would explain why free prequarks are not seen in experiments. An idea of this kind was first proposed by 't Hooft, who studied some of its mathematical implications but also expressed doubt that nature actually follows such a path.

The typical radius of hypercolor confinement must be less than 10^{-16} centimeter. Only when matter is probed at distances smaller than this would it be possible to see the hypothetical prequarks and their hypercolors. At a range of 10^{-14} or 10^{-15} centimeter hypercolor effectively disappears; the only objects visible at this scale of resolution (the quarks and leptons) are neutral with respect to hypercolor. At a range of 10^{-13} centimeter ordinary color likewise fades away, and the world seems to be made up entirely of objects that lack both color and hypercolor: protons, neutrons, electrons and so on.

The notion of hypercolor is well suited to a variety of prequark models, including the rishon model. In addition to their electric charge and color the rishons are assumed to have hypercolor and the antirishons to have antihypercolor. Only combinations of three rishons or three antirishons are allowed because only those combinations are neutral with respect to hypercolor. A mixed three-particle system, such as $TT\bar{T}$, cannot exist because it would not be hypercolorless. The assignment of hypercolors thereby explains the rule for forming composite rishon systems. Similar rules apply in other hypercolor-based prequark models.

If the aim of a prequark model is to simplify the understanding of nature, postulating a new basic force does not seem very helpful. In the case of hypercolor, however, there may be some compensation. Consider the neutrino: it has neither electric charge nor color, only weak charge. According to the standard model, two neutrinos can act on each other only through the short-range weak force. If neutrinos are composites of hypercolored prequarks, however, there could be an additional source of interactions between neutrinos. When two neutrinos are far apart, there are practically no hypercolor forces between them, but when they are at close range, the hypercolored prequarks inside one neutrino are able to 'see' the inner hypercolors of the other one. Complicated short-range attractions and repulsions are the result. The mechanism, of course, is exactly the same as the one that explains the molecular force as a residue of the electromagnetic force and the strong force as a residue of the color force.

The conclusion may also be the same. Seiberg and I, and independently Greenberg and Sucher, were the first to suggest that the short-range weak force may actually be a residual effect of the hypercolor force. According to this hypothesis, the weak bosons W^+, W^- and Z^0 must also be composite objects, presumably made up of certain combinations of the same prequarks that compose the quarks and leptons. If this idea is confirmed, the list of fundamental forces will still have four entries: gravitation, electromagnetism, color and hypercolor. It should be noted, however, that all these forces are long-range ones; the short-range molecular, strong and weak forces will have lost their fundamental status.

For now hypercolor remains a conjecture, and so does the notion of explaining the weak force as a residue of the hypercolor force. It may yet turn out that the weak force is fundamental. A careful measurement of the mass, lifetime and other properties of the weak bosons should provide important clues in this matter.

Hypercolor is not the only candidate for a prequark binding force. Another interesting possibility was suggested by Pati, Salam and Strathdee. Instead of introducing a new hypercolor force, they borrowed an idea that has long been familiar, namely the magnetic force, and adapted it to a new purpose. An ordinary magnet invariably has two poles, which can be thought of as opposite magnetic charges. For 50 years there have been theoretical reasons for supposing there could also be isolated magnetic charges, or monopoles. Pati, Salam and Strathdee have argued that the prequarks could be particles with charges resembling both magnetic and electric charges. If they are, the forces binding them may be of a new and interesting origin.

None of the ideas I have just described constitutes a theory of prequark dynamics. Indeed, there is a serious impediment to the formulation of such a theory; it is the requirement that the prequarks be exceedingly small. The most stringent limit on their size is set indirectly by measurements of the magnetic moment of the electron, which agree with the calculations of quantum electrodynamics to an accuracy of 10 significant digits. In the calculations it is assumed that the electron is pointlike; if it had any spatial extension or internal structure, the measured value would differ from the calculated one. Evidently any such discrepancy can at most affect the 11th digit of the result. It is this constraint that implies the characteristic distance scale of the electron's internal structure must be less than 10^{-16}

centimeter. Roughly speaking, that is the maximum radius of an electron, and any prequarks must stay within it. If they strayed any wider, their presence would already have been detected.

Why should the small size of the electron inhibit speculation about its internal structure? The uncertainty principle establishes a reciprocal relation between the size of a composite system and the kinetic energy

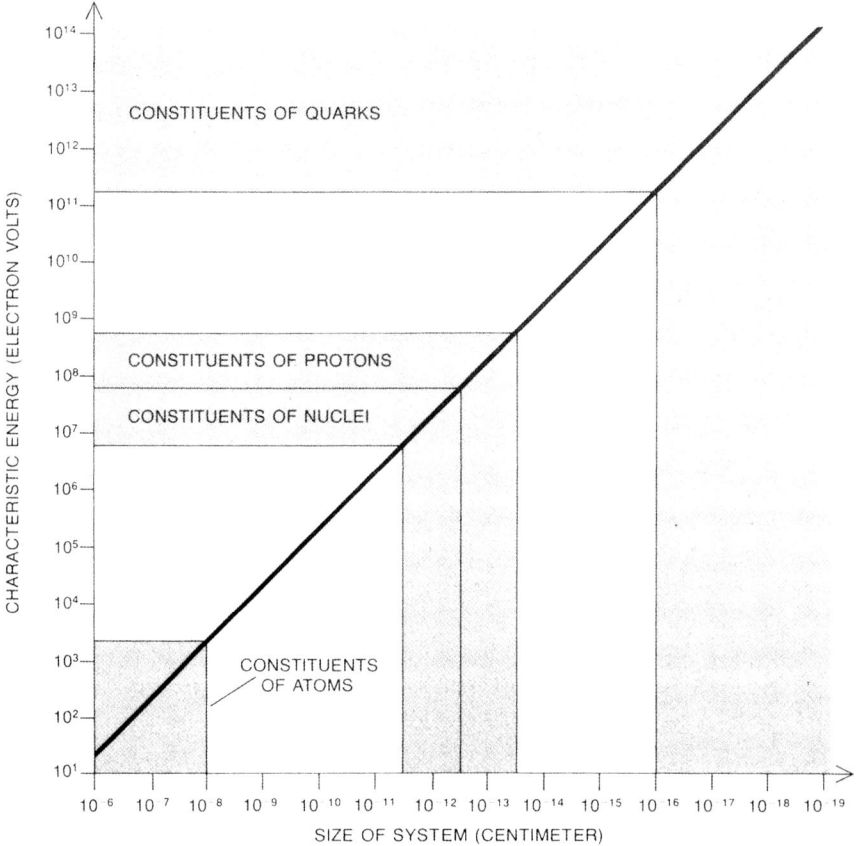

Figure 9. Size and energy have a reciprocal relation in the quantum theory, indicating that the constituents of composite quarks and leptons must have an exceedingly high kinetic energy. The size of an atom implies that its constituents can have energies ranging from a few electron volts to a few thousand. (An electron volt is the energy gained by an electron accelerated through a potential difference of one volt.) In a nucleus the protons and neutrons move with an energy of several million electron volts, and in a proton or a neutron the quarks have energies of several hundred million electron volts. Any constituents of quarks and leptons must be confined to a radius of less than 10^{-16} centimeter, and possibly much less. As a result the kinetic energy of the hypothetical prequarks can be no less than a few hundred billion electron volts.

of any components moving inside it. The smaller the system, the larger the kinetic energy of the constituents. It follows that the prequarks must have enormous energy: more than $100\,\mathrm{GeV}$ (100 billion electron volts), and possibly much more. (One electron volt is the energy acquired by an electron when it is accelerated through a potential difference of one volt.) Because mass is fundamentally equivalent to energy, it can be measured in the same system of units. The mass of the electron, for example, is equivalent to $0.0005\,\mathrm{GeV}$. There is a paradox here, which I call the energy mismatch: The

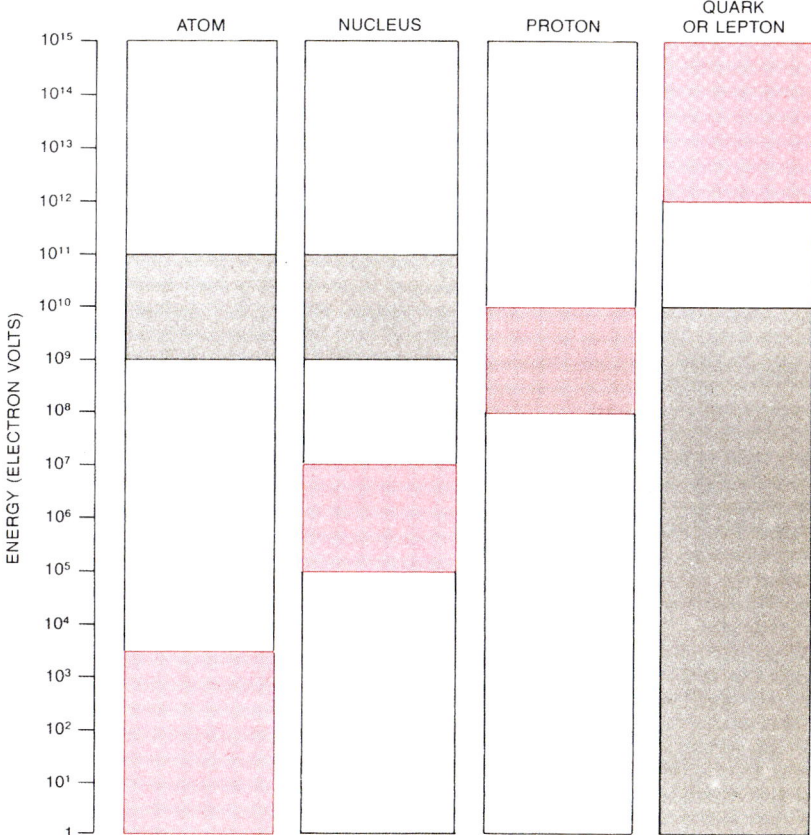

Figure 10. Mismatch of energy and mass makes it difficult to devise a theory of how prequarks might move and interact. In an atom or a nucleus the kinetic energy of the constituents (*color*) is much less than the total mass of the system (*gray*). In a proton the two quantities are of comparable magnitude. In a composite quark, however, the energy of the prequarks greatly exceeds the total mass. Indeed, compared with the kinetic energy, the mass is essentially zero. Somehow virtually all mass is canceled, a development that is unlikely to be accidental.

mass of the composite system (if it is indeed composite) is much smaller than the energy of its constituents.

The oddity of the situation can be illuminated by considering the relations of mass and kinetic energy in other composite systems. In an atom the kinetic energy of a typical electron is smaller than the mass of the atom by many orders of magnitude. In hydrogen, for example, the ratio is roughly one part in 100 million. The energy needed to change the orbit of the electron and thereby put the atom into an excited state is likewise a negligible fraction of the atomic mass. In a nucleus the kinetic energy of the protons and neutrons is also small compared with the nuclear mass, but it is not completely negligible. The motion of the particles gives them an energy equivalent to about 1 percent of the system's mass. The energy needed to create an excited state is also about 1 percent of the mass.

With the proton and its quark constituents the energy-mass relation begins to get curious. From the effective radius of the proton the typical energy of its component quarks can be calculated; it turns out to be comparable to the mass of the proton itself, which is a little less than 1 GeV. The energy that must be invested to create an excited state of the quark system is of the same order of magnitude: The hadrons identified as excited states of the proton exceed it in mass by from 30 to 100 percent. Nevertheless, the ratio of kinetic energy to total mass is still in the range that seems intuitively reasonable. Suppose one knew only the radius of the proton, and hence the typical energy of whatever happens to be inside it, and one were asked to guess the proton's mass. Since the energy of the constituents is generally a few hundred million electron volts, one would surely guess that the total mass of the system is at least of the same order of magnitude and possibly greater. The guess would be correct.

For the atom, the nucleus and the proton, then, the mass of the system is at least as large as the kinetic energy of the constituents and in some cases is much larger. If quarks and leptons are composite, however, the relation of energy to mass must be quite different. Since the prequarks have energies well above 100 GeV, one would guess that they would form composites with masses of hundreds of GeV or more. Actually the known quarks and leptons have masses that are much smaller; in the case of the electron and the neutrinos the mass is smaller by at least six orders of magnitude. The whole is much less than the sum of its parts.

The high energy of the prequarks is also what spoils the idea of viewing the higher generations of quarks and leptons as excited states of the same set of prequarks that form the first-generation particles. As in the other composite systems, the energy needed to change the orbits of the prequarks should be of the same order of magnitude as the kinetic energy of the constituents. One would therefore expect the successive generations to differ in mass by hundreds of GeV, whereas the actual mass differences are as small as 0.1 GeV.

At this point one might well adopt the view that the energy mismatch cannot be accepted, indeed that it simply demonstrates the elementary and structureless nature of the quarks and leptons. Many physicists hold this view. The energy mismatch, however, contradicts no basic law of physics, and I would argue that the circumstantial evidence for quark and lepton compositeness is sufficiently persuasive to warrant further investigation.

What is peculiar about the quark and lepton masses is not merely that they are small but that they are virtually zero when measured on the energy scale defined by their constituents' energy. In other composite systems a small amount of mass is 'lost' by being converted into the binding energy of the system. The total mass of a hydrogen atom, for example, is slightly less than that of an isolated proton and electron; the difference is equal to the binding energy. In a nucleus this 'mass defect' can reach a few percent of the total mass. In a quark or a lepton, it seems, the entire mass of the system is canceled almost exactly. Such a 'miraculous' cancellation is certainly not impossible, but it seems most unlikely to happen by accident. Similar large cancellations are known elsewhere in physics, and they have always been found to result from some symmetry principle or conservation law. If there is to be any hope of constructing a theory of prequark dynamics, it is essential to find such a symmetry in this case.

There is a likely candidate: chiral symmetry, or chirality. The name is derived from the Greek word for hand, and the symmetry has to do with handedness, the property defined by a particle's spin and direction of motion. Like other symmetries of nature, chiral symmetry has a conservation law associated with it, which gives the clearest account of what the symmetry means. The law states that the total number of right-handed particles and the total number of left-handed ones can never change.

In the ordinary world of protons, electrons and similar particles handedness or chirality clearly is not conserved. A violation of the conservation law can be demonstrated by a simple thought experiment. Imagine that an observer is moving in a straight line when he is overtaken by an electron. As the electron recedes from him he notes that its spin and direction of motion are related by a right-hand rule. Now suppose the observer speeds up, so

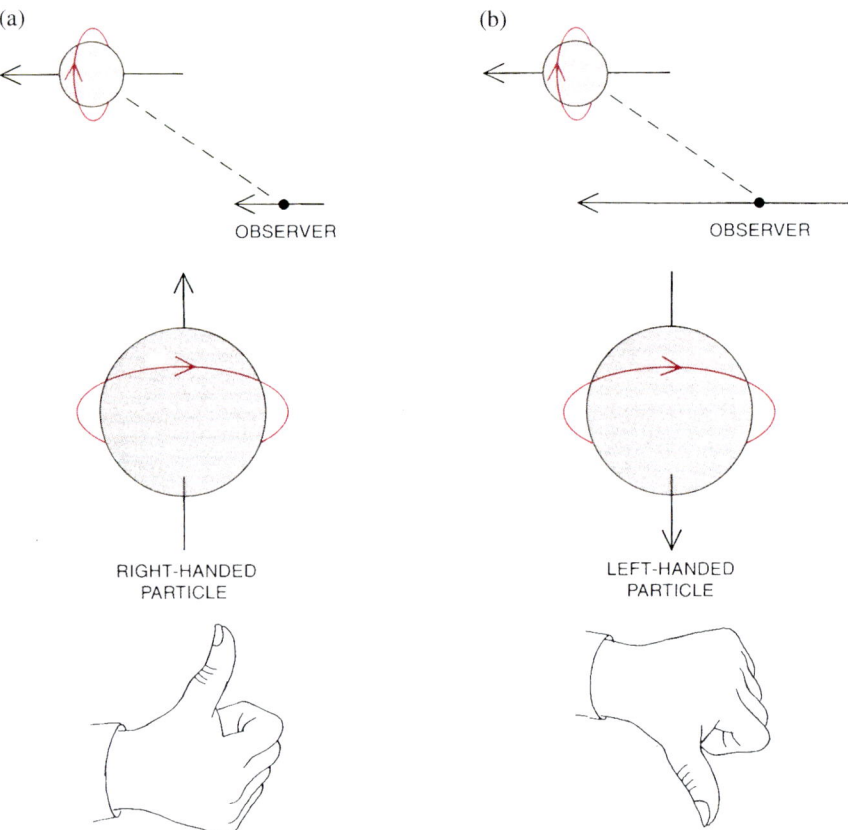

Figure 11. Chiral symmetry offers a possible explanation of the 'miraculous' cancelation of mass in quarks and leptons. Chirality, or handedness, decribes the relation of a particle's spin angular momentum to its direction of motion. Suppose an observer is overtaken by a faster-moving electron (a). From the observer's point of view the electron obeys a right-hand rule: When the fingers of the right hand curl in the same direction as the spin, the thumb gives the direction of motion. If the observer speeds up, however, so that he overtakes the electron (b), the handedness of the particle changes. In the observer's frame of reference the electron is now approaching instead of receding, but its spin direction has not changed; as a result its motion is described by a left-hand rule. Chirality, therefore, is not conserved. There is one kind of particle to which this argument cannot be applied, namely a massless particle, which must always move with the speed of light. No observer can move faster than a massless particle, and so its handedness is an invariant property. If a theory of prequarks had a chiral symmetry, in which handedness must be conserved, the low mass of the quarks and leptons might not be accidental. They would have to be virtually massless for the chiral symmetry to be maintained.

that he is overtaking the electron. In the observer's frame of reference the electron seems to be approaching; in other words, it has reversed direction. Because its spin has not changed, however, it has become a left-handed particle.

There is one kind of particle to which this thought experiment cannot be applied: a massless particle. Because a massless particle must always move with the speed of light, no observer can ever go faster. As a result the handedness of a massless particle is an invariant property, independent of the observer's frame of reference. Furthermore, it can be shown that none of the known forces of nature (those mediated by the photon, the gluons and the weak bosons) can alter the handedness of a particle. Thus if the world were made up exclusively of massless particles, the world could be said to have chiral symmetry.

Chiral symmetry is the root of an idea that might conceivably account for the small mass of the quarks and leptons. The argument runs as follows. If the prequarks are massless particles, if they have a spin of $1/2$ and if they interact with one another only through the exchange of gauge bosons, any theory describing their motion is guaranteed to have a chiral symmetry. If the massless prequarks then bind together to form composite spin-$1/2$ objects (namely the quarks and leptons), the chiral symmetry might ensure that the composite particles also remain massless compared with the huge energy of the prequarks inside them. Hence the small mass of the quarks and leptons is not an accident. They must be essentially massless with respect to the energy of their constituents if the chiral symmetry of the theory is to be maintained.

The crucial step in this argument is the one extending the chiral symmetry from a world of massless prequarks to one made up of composite quarks and leptons. It is essential that the symmetry of the original physical system survive in and be respected by the composite states formed out of the massless constituents. It may seem self-evident that if a theory is symmetrical in some sense, the physical systems described by the theory must exhibit that symmetry; actually, however, the spontaneous breaking of symmetries is commonplace. A familiar example is the roulette wheel. A physical theory of the roulette wheel would show it is completely symmetrical in the sense that each slot is equivalent to any other slot. The physical system formed by putting a ball in the roulette wheel, however, is decidedly asymmetrical: the ball invariably comes to rest in just one slot.

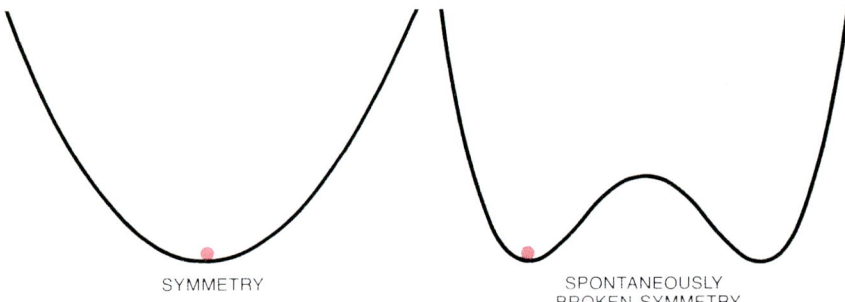

SYMMETRY   SPONTANEOUSLY
                                  BROKEN SYMMETRY

Figure 12. Spontaneous symmetry breaking is a mechanism that could spoil a prequark theory even if it has a chiral symmetry. Both of the physical systems shown here — a simple trough and a trough with a bump in the bottom — can be described as symmetrical in the sense that exchanging left and right leaves the system unaltered. For the simple trough the system remains symmetrical when a ball is put in the trough; the ball comes to rest in the center, so that exchanging left and right still has no effect. In the trough with a bump, however, the ball takes up a position on one side or the other, and the symmetry is inevitably broken. Similarly, a prequark theory that has a chiral symmetry might nonetheless give rise to composite systems that do not observe the symmetry. Showing that a chiral symmetry can definitely remain unbroken is currently the principal challenge in formulating a theory of how prequarks move.

In the standard model it is the spontaneous breaking of a symmetry that makes the three weak bosons massive and leaves the photon massless. The theory that describes these gauge bosons is symmetrical, and in it the four bosons are essentially indistinguishable, but because of the symmetry breaking the physical states actually observed are quite different. Chiral symmetries are notoriously susceptible to symmetry breaking. Whether the chiral symmetry of prequarks breaks or not when the prequarks form composite objects can be determined only with a detailed understanding of the forces acting on the prequarks. For now that understanding does not exist. In certain models it can be shown that a chiral symmetry does exist but is definitely broken. No one has yet succeeded in constructing a composite model of quarks and leptons in which a chiral symmetry is known to remain unbroken. Neither the preon model nor the rishon model succeeds in solving the problem. The task is probably the most difficult one facing those attempting to demonstrate that quarks and leptons are composite.

If a consistent prequark theory can be worked out, it will still have to pass the test of experiment. First, it is important to establish in the laboratory whether or not quarks and leptons have any internal structure at all. If they do, experiments might then begin to discriminate among the various models. The experiments will have to penetrate the unknown realm of distances smaller than 10^{-16} centimeter and energies higher than $100\,\text{GeV}$. There are two basic ways to explore this region: by doing experiments with particles accelerated to very high energy and by making precise measurements of low-energy quantities that depend on the physics of events at very small distances.

Experiments of the first kind include the investigation of the weak bosons and the search for the Higgs particles of the standard model. When such particles can be made in sufficient numbers, a careful look at their properties should reveal much about the physics of very small distances. New accelerators now being planned or built in the U.S., Europe and Japan are expected to yield detailed information about the weak bosons and will also continue the ongoing investigation of the quarks and leptons themselves.

Equally interesting are the high-precision, low-energy experiments. One of these is the search for the decay of the proton, a particle that is known to have an average lifetime of at least 10^{30} years. Several experiments are now monitoring large quantities of matter, incorporating substantially more than 10^{30} protons, in an attempt to detect the signals emitted when a proton disintegrates. None of the forces of the standard model can induce such an event, but none of the rules of the standard model absolutely forbids it. Both the grand unified theories and the prequark models, on the other hand, include mechanisms that could convert a proton into other particles that would ultimately leave behind only leptons and photons. If the decay is detected, its rate and the pattern of decay products could offer an important glimpse beyond the standard model.

There is similar interest in the hypothetical process in which a muon emits a photon and is thereby converted into an electron. Again none of the forces of the standard model can bring about an event of this kind, but again too no fundamental law forbids it. Some of the composite models allow the transition and others do not, so that a search for the process might offer a means of choosing among the models. Experiments done up to now put a limit of less than one in 10 billion on the probability that any given muon will decay in this way. Detection of such events and a determination of their rate might illuminate the mysterious distinction between the generations.

A third class of precision experiments are those that continue to refine the measurement of the magnetic moment of the electron and of the muon.

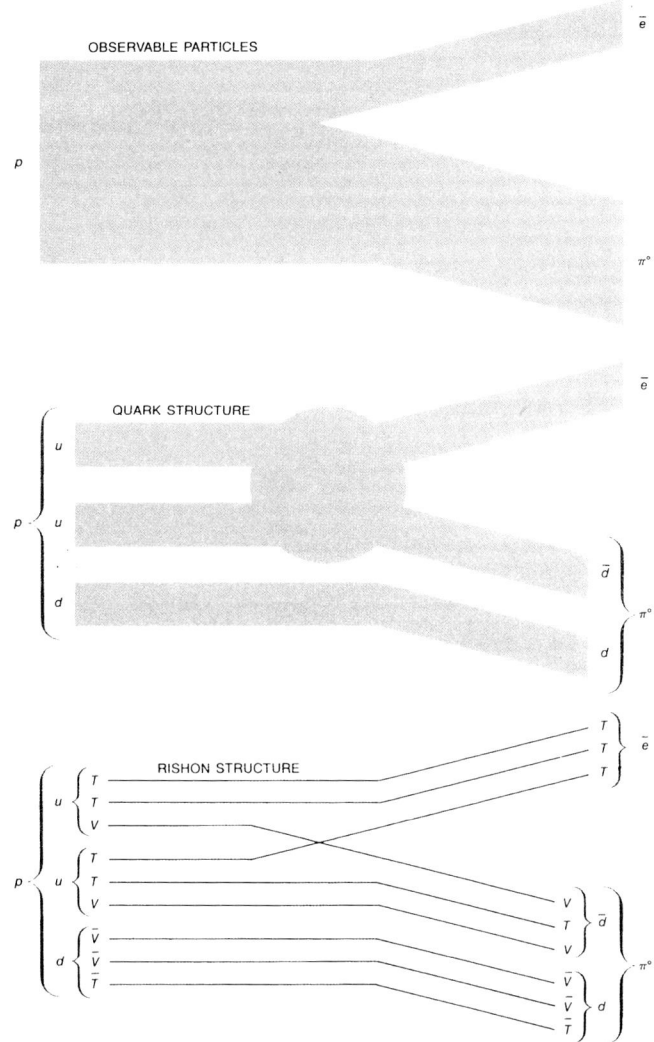

Figure 13. Decay of the proton is a conjectured event that might be intepreted as experimental evidence for a grand unified theory or for a model of quark substructure. In one form of decay the proton would be observed to disintegrate into a positron ($\bar{e}$) and a neutral pion (π^0). The event can be understood in terms of the proton's quark constituents: An interaction of the two u quarks converts one of them into a positron and the other into a $\bar{d}$ antiquark; the latter combines with the remaining d quark of the proton to form the neutral pion. Grand unified theories suggest that the interaction of the u quarks is mediated by a new force of nature. The rishon model provides an alternative explanation: The two u quarks merely exchange a T and a V rishon.

Further improvements can be expected both in experimental accuracy and in the associated calculations of quantum electrodynamics. If the results continue to agree with the predictions of the standard model, the limit on the possible size of any quark and lepton substructure will become remoter. If a discrepancy between theory and experiment is detected, it will represent a strong hint that quarks and leptons are not elementary.

It may well be a decade or two before the next level in the structure of matter comes clearly into view (if, again, there is another level). What is needed is a sound theoretical picture, one that is self-consistent, that agrees with all experiments and that is simple enough to explain all the features of the standard model in terms of a few principles and a few fundamental particles and forces. The correct picture, whether it is a grand unified theory or a composite model of the quarks and leptons, may already exist in some embryonic form. On the other hand, it is also possible the correct theory will emerge only from some totally new idea. In the words of Niels Bohr, it may be that our present ideas 'are not sufficiently crazy to be correct.'

Editor's note: In the span of time since this article was written, experimental discoveries have altered some statements made in it. The top quark, the tau neutrino and the weak interaction bosons have been discovered (but the Higgs particle is still at large). And there is no longer a possibility that any of the neutrino species is massless, as attested by the phenomenon of neutrino oscillations.

Prof. Haim Harari delivered the Albert Einstein Memorial Lecture in 1983.

Beautiful Theories

Steven Weinberg

When on some gilded cloud or flowre
My gazing soul would dwell an houre,
And in those weaker glories spy
Some shadows of eternity.

Henry Vaughn, *The Retreat*

In 1974 Paul Dirac came to Harvard to speak about his historic work as one of the founders of modern quantum electrodynamics. Toward the end of his talk he addressed himself to our graduate students, and advised them to be concerned only with the beauty of their equations, not with what the equations mean. It was not good advice for students, but the search for beauty in physics[1] was a theme that ran throughout Dirac's work, and indeed through much of the history of physics.

Some of the talk about the importance of beauty in science has been little more than gushing. I do not propose to use this talk just to say more nice things about beauty. Rather, I want to focus more closely on the nature of beauty in physical theories, on why our sense of beauty is sometimes a useful guide and sometimes not, and on how the usefulness of our sense of beauty is a sign of our progress toward a final theory.

*This article is reprinted from *Dreams of a Final Theory* by Steven Weinberg, copyright ©1992 by Steven Weinberg. Used by permission of Random House, Inc., throughout the world excluding the UK, and of Vintage, an imprint of The Random House Group, Ltd., in the UK.
[1]The astrophysicist Subrahmanyan Chandresekhar has written movingly of the role of beauty in science, in *Truth and Beauty: Aesthetics and Motivations in Science*, Chicago: University of Chicago Press, 1987, and *Bulletin of the American Academy of Arts and Sciences*, XLIII, No. 3 (December 1989), p. 14.

A physicist who says that a theory is beautiful does not mean quite the same thing that would be meant in saying that a particular painting or a piece of music or poetry is beautiful. It is not merely a personal expression of aesthetic pleasure; it is much closer to what a horse trainer means when he looks at a racehorse and says that it is a beautiful horse. The horse trainer is of course expressing a personal opinion, but it is an opinion about an objective fact: that, on the basis of judgments that the trainer could not easily put into words, this is the kind of horse that wins races.

Of course, different horse trainers may judge a horse differently. That is what makes horse racing. But the horse trainer's aesthetic sense is a means to an objective end — the end of selecting horses that will win races. The physicist's sense of beauty is also supposed to serve a purpose — it is supposed to help the physicist select ideas that will help us to explain nature. Physicists, just as horse trainers, may be right or wrong in their judgments, but they are not merely enjoying themselves. They often *are* enjoying themselves, but that is not the whole purpose of their aesthetic judgements.

This comparison raises more questions than it answers. First, what *is* a beautiful theory? What are the characteristics of physical theories that give us a sense of beauty? A more difficult question: Why does the physicist's sense of beauty work, when it does work? The history of physics provides many illustrations of the rather spooky fact that something as personal and subjective as our sense of beauty helps us not only to invent physical theories but even to judge the validity of theories. Why are we blessed with such aesthetic insight? The effort to answer the question raises another question that is even more difficult, although perhaps it sounds trivial: What is it that the physicist wants to accomplish?

What is a beautiful theory? The curator of a large American art museum once became indignant at my use of the word 'beauty' in connection with physics. He said that in his line of work professionals have stopped using this word because they realize how impossible it is to define. Long ago the physicist and mathematician Henri Poincaré admitted that 'it may be very hard to define mathematical beauty, but that is just as true of beauty of all kinds.'

I will not try to define beauty, any more than I would try to define love or fear. You do not define these things; you know them when you feel them. Later, after the fact, you may sometimes be able to say a little to describe them, as I will try to do here.

By the beauty of a physical theory, I certainly do not mean merely the mechanical beauty of its symbols on the printed page. The metaphysical poet Thomas Traherne took pains that his poems should make pretty patterns on the page, but this is no part of the business of physics. I should also distinguish the sort of beauty I am talking about here from the quality that mathematicians and physicists sometimes call elegance. An elegant proof or calculation is one that achieves a powerful result with a minimum of irrelevant complication. It is not important for the beauty of a theory that its equations should have elegant solutions. The equations of general relativity are notoriously difficult to solve except in the simplest situations, but this does not detract from the beauty of the theory itself. Einstein has been quoted as saying that scientists should leave elegance to tailors.

Simplicity is part of what I mean by beauty, but it is a simplicity of ideas, not simplicity of a mechanical sort that can be measured by counting equations or symbols. Both Einstein's and Newton's theories of gravitation involve equations that tell us the gravitational forces produced by any given amount of matter. In Newton's theory there are three of these equations (corresponding to the three dimensions of space) — in Einstein's theory there are fourteen.[2] In itself, this cannot be counted as an aesthetic advantage of Newton's theory over Einstein's. And in fact it is Einstein's theory that is more beautiful, in part because of the simplicity of his central idea about the equivalence of gravitation and inertia. That is a judgment on which scientists have generally agreed, and it was largely responsible for the early acceptance of Einstein's theory.

There is another quality besides simplicity that can make a physical theory beautiful — it is the sense of inevitability that the theory may give us. In listening to a piece of music or hearing a sonnet one sometimes feels an intense aesthetic pleasure at the sense that nothing in the work could be changed, that there is not one note or one word that you would want to have different. In Raphael's *Holy Family* the placement of every figure on the canvas is perfect. This may not be of all paintings in the world your favorite, but as you look at that painting there is nothing that you would want Raphael to have done differently. The same is partly true (it is never more than partly true) of general relativity. Once you know the general physical principles adopted by Einstein, you understand that there is no other significantly different theory of gravitation to which Einstein could have been led. As Einstein said of general relativity, 'The chief attraction of the theory lies in its logical completeness. If a single one of the conclusions

---

[2]I am referring to the ten field equations plus the four equations of motion.

drawn from it proves wrong, it must be given up; to modify it without destroying the whole structure seems to be impossible.'[3]

This is less true of Newton's theory of gravitation. Newton could have supposed that the gravitational force decreases with the inverse cube of distance rather than the inverse square if that is what the astronomical data had demanded, but Einstein could not have incorporated an inverse-cube law in his theory without scrapping its conceptual basis. Thus Einstein's fourteen equations have an inevitability and hence beauty that Newton's three equations lack. I think that this is what Einstein meant when he referred to the side of the equations that involve the gravitational field in his general theory of relativity as beautiful, as if made of marble, in contrast with the other side of the equations, referring to matter, which he said were still ugly, as if made of mere wood. The way that the gravitational field enters Einstein's equations is almost inevitable, but nothing in general relativity explained why matter takes the form it does.

The same sense of inevitability can be found (again, only in part) in our modern standard model of the strong and electroweak forces that act on elementary particles. There is one common feature that gives general relativity and the standard model most of their sense of inevitability and simplicity: They obey *principles of symmetry.*

A symmetry principle is simply a statement that something looks the same from certain different points of view. Of all such symmetries, the simplest is the approximate bilateral symmetry of the human face. Because there is little difference between the two sides of your face, it looks the same whether viewed directly or when left and right are reversed, as when you look in a mirror. It is almost a cliché of filmmaking to let the audience realize suddenly that the actor's face they have been watching has been seen in a mirror; the surprise would be spoiled if people had two eyes on the same side of the face like flounders, and always on the same side.

Some things have more extensive symmetries than the human face. A cube looks the same when viewed from six different directions, all at right angles to each other, as well as when left and right are reversed. Perfect crystals look the same not only when viewed from various different directions but also when we shift our positions within the crystal by certain amounts in various directions. A sphere looks the same from any direction. Empty space looks the same from all directions and all positions.

[3]Quoted by G. Holton, 'Constructing a Theory: Einstein's Model,' *American Scholar*, 48 (Summer 1979), p. 323.

Symmetries like these have amused and intrigued artists and scientists for millennia but did not really play a central role in science. We know many things about salt, and the fact that it is a cubic crystal and therefore looks the same from six different points of view does not rank among the most important. Certainly bilateral symmetry is not the most interesting thing about a human face. The symmetries that are really important in nature are not the symmetries of *things*, but the symmetries of *laws*.

A symmetry of the laws of nature is a statement that when we make certain changes in the point of view from which we observe natural phenomena, the laws of nature we discover do not change. Such symmetries are often called principles of *invariance*. For instance, the laws of nature that we discover take the same form however our laboratories are oriented; it makes no difference whether we measure directions relative to north or northeast or upward or any other direction. This was not so obvious to ancient or medieval natural philosophers; in everyday life there certainly seems to be a difference between up and down and horizontal directions. Only with the birth of modern science in the seventeenth century did it become clear that down seems different from up or north only because below us there happens to be a large mass, the earth, and not (as Aristotle thought) because the natural place of heavy or light things is downward or upward. Note that this symmetry does not say that up is the same as down; observers who measure distances upward or downward from the earth's surface report different descriptions of events such as the fall of an apple, but they discover the same laws, such as the law that apples are attracted by large masses like the earth.

The laws of nature also take the same form wherever our laboratories are located; it makes no difference to our results whether we do our experiments in Texas or Switzerland or on some planet on the other side of the galaxy. The laws of nature take the same form however we set our clocks; it makes no difference whether we date events from the Hegira or the birth of Christ or the beginning of the universe. This does not mean that nothing changes with time or that Texas is just the same as Switzerland, only that the laws discovered at different times and in different places are the same. If it were not for these symmetries the work of science would have to be redone in every new laboratory and in every passing moment.

Any symmetry principle is at the same time a principle of simplicity. If the laws of nature did distinguish among directions like up or down or north, then we would have to put something into our equations to keep track of the orientation of our laboratories, and they would be correspondingly less simple. Indeed, the very notation that is used by mathematicians and physicists to make our equations look as simple and compact as possible has built into it an assumption that all directions in space are equivalent.

Important as these symmetries of the laws of nature are in classical physics, their importance is even greater in quantum mechanics. Consider, what makes one electron different from another? Only its energy, its momentum and its spin; aside from these properties, every electron in the universe is the same as every other. All these properties of an electron are simply quantities that characterize the way that the quantum-mechanical wave function of the electron responds to symmetry transformations: to changing the way we set our clocks or the location or orientation of our laboratory.[4] Matter thus loses its central role in physics: All that is left is principles of symmetry and various ways that wave functions can behave under symmetry transformations.

---

[4]E.g., the frequency with which the wave function of any system in a state of definite energy oscillates is given by the energy divided by a constant of nature known as Planck's constant. This system appears much the same to two observers who have set their watches differently by one second, but, if they both observe the system when the hands on their watches both point precisely to twelve noon, they observe that the oscillation is at a different phase; because their watches are set differently they are really observing the system at different times, so that one observer may, e.g., see a crest in the wave, while the other sees a trough. Specifically, the phase is different by the number of cycles (or parts of cycles) that occur in one second; in other words, by the frequency of the oscillation in cycles per second, and hence by the energy divided by Planck's constant. In today's quantum mechanics, we *define* the energy of any system as the change in phase (in cycles or parts of cycles) of the wave function of the system at a given *clock* time when we shift the way our clocks are set by one second. Planck's constant gets into the act only because energy is historically measured in units like calories or kilowatt hours or electron volts that were adopted before the advent of quantum mechanics; Planck's constant simply provides the conversion factor between these older systems of units and the natural quantum-mechanical unit of energy, which is cycles per second. It can be shown that energy defined in this way has all the properties that we normally associate with energy, including its conservation; indeed, the invariance of the laws of nature under the symmetry transformation of re-setting our watches is *why* there is such a thing as energy. In much the same way, the component of the momentum of any system in any particular direction is defined as the change of phase of the wave function when we shift the point from which positions are measured by one centimeter in that direction, again times Planck's constant. The amount of spin of a system around any axis is defined as the change of the phase of the wave function when we rotate the frame of reference we use for measuring directions around that axis by one full turn, times Planck's constant. From this point of view, momentum and spin are what they are because of the symmetry of the laws of nature under changes in the frame of reference that we use to measure positions or directions in space. (In listing the properties of electrons I do not include position, because position and momentum are complementary properties; we can describe the state of an electron in terms of its position *or* its momentum, but not both together.)

There are symmetries of space-time that are less obvious than these simple translations or rotations. The laws of nature also appear to take the same form to observers moving at different constant velocities; it makes no difference whether we do our experiments here in the solar system, whizzing around the center of the galaxy at hundreds of kilometers per second, or in a distant galaxy rushing away from our own at tens of thousands of kilometers per second. This last symmetry principle is sometimes called the principle of relativity. There is a widespread impression that this principle was invented by Einstein, but there was also a principle of relativity in Newton's theory of mechanics; the difference is only in the way that the speed of the observer affects observations of positions and times in the two theories. But Newton took his version of the principle of relativity for granted; Einstein explicitly designed his version of the principle of relativity to be consistent with an experimental fact, that the speed of light seems the same however the observer is moving. In this sense the emphasis on symmetry as a question of physics in Einstein's 1905 paper on special relativity marks the beginning of the modern attitude to symmetry principles.

The most important difference between the way that observations of space-time positions are affected by the motion of observers in Newton's and Einstein's physics is that in special relativity there is no absolute meaning to a statement that two distant events are simultaneous. One observer may see that two clocks strike noon at the same moment; another observer that is moving with respect to the first finds that one clock strikes noon before or after the other. This makes Newton's theory of gravitation or any similar theory of force inconsistent with special relativity. Newton's theory tells us that the gravitational force that the sun exerts on the earth at any one moment depends on where the sun's mass is at the same moment, but the same moment according to whom?

The natural way to avoid this problem is to abandon the old Newtonian idea of instantaneous action at a distance and to replace it with a picture of force as due to *fields*. In this picture the sun does not directly attract the earth; rather, it creates a field, called the gravitational field, that then exerts a force on the earth. This might seem like a distinction without a difference, but there is a crucial difference: When a solar flare erupts on the sun, it first affects the gravitational field only near the sun, after which the tiny change in the field propagates through space at the speed of light like ripples spreading out from where a pebble falls into water, only reaching the earth some eight minutes later. All observers moving at any constant velocity agree with this description, because in special relativity all such observers agree about the speed of light. In the same way, an electrically charged body creates a field, called the electromagnetic field, that exerts electric and magnetic forces on other charged bodies. When an electrically charged body is suddenly moved, the electromagnetic field is at first changed only near that body, and the changes in this field then propagate at the speed of light. In fact, in this case the changes in the electromagnetic field *are* what we know as light, though it is often light whose wavelength is so low or high that it is not visible to us.

In the context of prequantum physics Einstein's special theory of relativity fit in well with a dualistic view of nature: There are particles, like the electrons, protons, and neutrons in ordinary atoms, and there are fields, like the gravitational or the electromagnetic field. The advent of quantum mechanics led to a much more unified view. Quantum mechanically, the energy and momentum of a field like the electromagnetic field comes in bundles known as photons that behave exactly like particles, though like particles that happen to have no mass. Similarly, the energy and momentum in the gravitational field come in bundles called gravitons,[5] that also behave like particles of zero mass. In a large-scale field of force like the sun's gravitational field we do not notice individual gravitons, essentially because there are so many of them.

In 1929 Werner Heisenberg and Wolfgang Pauli (building on earlier work of Max Born, Heisenberg, Pascual Jordan and Eugene Wigner) explained in a pair of papers how massive particles like the electron could also be understood as bundles of energy and momenta in different sorts of fields, such as the electron field. Just as the electromagnetic force between two electrons is due in quantum mechanics to the exchange of photons, the force between photons and electrons is due to the exchange of electrons. The distinction between matter and force largely disappears; any particle can play the role of a test body on which forces act and by its exchange can produce other forces. Today it is generally accepted that the only way to combine the principles of special relativity and quantum mechanics is through the quantum theory of fields or something very like it. This is precisely the sort of logical rigidity that gives a really fundamental theory its beauty: Quantum mechanics and special relativity are nearly

---

[5]Gravitons have not been detected experimentally, but this is no surprise; calculations show that they interact so weakly that individual gravitons could not have been detected in any experiment yet performed. Nevertheless, there is no serious doubt of the existence of gravitons.

incompatible, and their reconciliation in quantum field theory imposes powerful restrictions on the ways that particles can interact with one another.

All the symmetries mentioned so far only limit the kinds of force and matter that a theory may contain — they do not in themselves *require* the existence of any particular type of matter or force. Symmetry principles have moved to a new level of importance in this century and especially in the last few decades: There are symmetry principles that dictate the very existence of all of the known forces of nature.

In general relativity the underlying principle of symmetry states that *all* frames of reference are equivalent: The laws of nature look the same not only to observers moving at any constant speed but to all observers, whatever the acceleration or rotation of their laboratories. Suppose we move our physical apparatus from the quiet of a university laboratory, and do our experiments on a steadily rotating merry-go-round. Instead of measuring directions relative to north, we would measure them with respect to the horses fixed to the rotating platform. At first sight the laws of nature will appear quite different. Observers on a rotating merry-go-round observe a centrifugal force that seems to pull loose objects to the outside of the merry-go-round. If they are born and grow up on the merry-go-round and do not know that they are on a rotating platform, they describe nature in terms of laws of mechanics that incorporate this centrifugal force, laws that appear quite different from those discovered by the rest of us.

The fact that the laws of nature seem to distinguish between stationary and rotating frames of reference bothered Isaac Newton and continued to trouble physicists in the following centuries. In the 1880s the Viennese physicist and philosopher Ernst Mach pointed the way toward a possible reinterpretation. Mach emphasized that there was something else besides centrifugal force that distinguishes the rotating merry-go-round and more conventional laboratories. From the point of view of an astronomer on the merry-go-round, the sun, stars, galaxies — indeed, the bulk of the matter of the universe — seems to be revolving around the zenith. You or I would say that this is because the merry-go-round is rotating, but an astronomer who grew up on the merry-go-round and naturally uses *it* as his frame of reference would insist that it is the rest of the universe that is spinning around him. Mach asked whether there was any way that this great apparent circulation of matter could be held responsible for centrifugal force. If so, then the laws of nature discovered on the merry-go-round might actually be the same as those found in more conventional laboratories; the apparent difference would simply arise from the different environment seen by observers in their different laboratories.

Mach's hint was picked up by Einstein and made concrete in his general theory of relativity. In general relativity there is indeed an influence exerted by the distant stars that creates the phenomenon of centrifugal force in a spinning merry-go-round: It is the force of gravity. Of course nothing like this happens in Newton's theory of gravitation, which deals only with a simple attraction between all masses. General relativity is more complicated; the circulation of the matter of the universe around the zenith seen by observers on the merry-go-round produces a field somewhat like the magnetic field produced by the circulation of electricity in the coils of an electromagnet. It is this 'gravitomagnetic' field that in the merry-go-round frame of reference produces the effects that in more conventional frames of reference are attributed to centrifugal force. The equations of general relativity, unlike those of Newtonian mechanics, are precisely the same in the merry-go-round laboratory and conventional laboratories; the difference between what is observed in these laboratories is entirely due to their different environment — a universe that revolves around the zenith, or one that does not. But, if gravitation did not exist, this reinterpretation of centrifugal force would be impossible, and the centrifugal force felt on a merry-go-round would allow us to distinguish between the merry-go-round and more conventional laboratories and would thus rule out any possible equivalence between laboratories that are rotating and those that are not. *Thus the symmetry among different frames of reference requires the existence of gravitation.*

The symmetry underlying the electroweak theory is a little more esoteric. It does not have to do with changes in our point of view in space and time but rather with changes in our point of view about the identity of the different types of elementary particle. Just as it is possible for a particle to be in a quantum mechanical state in which it is neither definitely here nor there, or spinning neither definitely clockwise nor counterclockwise, so also through the wonders of quantum mechanics it is possible to have a particle in a state in which it is neither definitely an electron nor definitely a neutrino until we measure some property that would distinguish the two, like the electric charge. In the electroweak theory the form of the laws of nature is unchanged if we replace electrons and neutrinos everywhere in our equations with such mixed states that are neither electrons nor neutrinos. Because various other types of particles interact with electrons and neutrinos, it is necessary at the same time to mix up families of these other particle types, such as up quarks with down quarks, as well as the photons with its siblings, the positively and negatively charged W particles and the neutral Z particles.[6] This is the symmetry that connects the electromagnetic forces, which are produced by an exchange of photons, with the weak nuclear forces that are produced by the exchange of the W and Z particles. The photon and the W and Z particles appear in the electroweak theory as bundles of the energy of four fields, fields that are required by this symmetry of the electroweak theory in much the same way that the gravitational field is required by the symmetries of general relativity.

Symmetries of the sort underlying the electroweak theory are called *internal symmetries*, because we can think of them as having to do with the intrinsic nature of the particles, rather than their position or motion. Internal symmetries are less familiar than those that act on ordinary space and time, such as those governing general relativity. You can think of each particle as carrying a little dial, with a pointer that points in directions marked 'electron' or 'neutrino' or 'photon' or 'W' or anywhere in between. The internal symmetry says that the laws of nature take the same form if we rotate the markings on these dials in certain ways.

Furthermore, for the sort of symmetry that governs the electroweak forces, we can rotate the dials differently for particles at different times or positions. This is much like the symmetry underlying general relativity, which allows us to rotate our laboratory not only by some fixed angle but also by an amount that increases with time, by placing it on a merry-go-round. The invariance of the laws of nature under a group of such position-dependent and time-dependent internal symmetry transformations is called a *local* symmetry (because the effect of the symmetry transformations

---

[6]Strictly speaking, it is only the left-handed states of the electron and neutrino and the up-and-down quarks that form these families. (By left-handed, I mean that the particle spins in the direction that your fingers curl if the thumb of your left hand is laid along the axis of rotation of the particle pointing in the particle's direction of motion.) This distinction between the families formed by left- and right-handed states is the origin of the fact that the weak nuclear forces do not respect the symmetry between right and left. (The asymmetry between right and left in the weak forces was proposed in 1956 by the theorists T.D. Lee and C.N. Yang. It was verified by experiments on nuclear beta decay by C.S. Wu in collaboration with a group at the National Bureau of Standards in Washington, and in experiments of the decay of the pi meson by R.L. Garwin, L. Lederman, and M. Weinrich and by J. Friedman and V. Telegdi.) We still do not know why it is only the left-handed electrons and neutrinos and quarks that form these families; this is a challenge for theories that aim at going beyond our standard model of elementary particles.

depends on location in space and time) or a *gauge* symmetry (for reasons that are purely historical).[7] It is the local symmetry between different frames of reference in space and time that makes gravitation necessary, and in much the same way it is a second local symmetry between electrons and neutrinos (and between up quarks and down quarks and so on) that makes the existence of the photon, W, and Z fields necessary.

There is yet a third exact local symmetry associated with the internal property of quarks that is fancifully known as *color*.[8] We have seen that there are quarks of various types, like the up and down quarks that make up the protons and neutrons found in all ordinary atomic nuclei. In addition, each of these types of quarks comes in three different 'colors,' which physicists (at least in the United States) often call red, white and blue. Of course, this has nothing to do with ordinary color; it is merely a label used to distinguish different subvarieties of quark. As far as we know, there is an exact symmetry in nature among the different colors; the force between a red quark and a white quark is the same as between a white quark and a blue quark, and the force between two red quarks is the same as between two blue quarks. But this symmetry goes beyond mere interchanges of colors. In quantum mechanics we can consider states of a single quark that are neither definitely red nor definitely white nor definitely blue. The laws of nature take precisely the same form if we replace red, white and blue quarks with quarks in three suitable mixed states (e.g., purple, pink and lavender). Again in analogy with general relativity, the fact that the laws of nature are unaffected even if the mixtures vary from place to place and time to time makes it necessary to include a family of fields in the theory that interact with quarks, analogous to the gravitational field. There are eight of these fields; they are known as gluon fields because the strong forces they produce glue the quarks together inside the proton and neutron. Our

---

[7]In 1918 the mathematician Hermann Weyl proposed that the symmetry of general relativity under space-time-dependent changes of position or orientation should be supplemented by a symmetry under space-time-dependent changes in the way one measures (or 'gauges') distances and times. This symmetry principle was soon abandoned by physicists (though versions of it crop up now and then in speculative theories) but it is mathematically very similar to an internal symmetry of the electrodynamics, which therefore came to be called gauge invariance. Then, when a more complicated sort of local internal symmetry was introduced in 1954 by C.N. Yang and R.L. Mills (in an attempt to account for the strong nuclear force), it, too, was called a gauge symmetry.

[8]Various versions of the attribute of quarks known as color were suggested by O.W. Greenberg; M.Y. Han and Y. Nambu; and W.A. Bardeen, H. Fritzsch and M. Gell-Mann.

modern theory of these forces, *quantum chromodynamics*, is nothing but the theory of quarks and gluons that respects this local color symmetry. The standard model of elementary particles consists of the electroweak theory combined with quantum chromodynamics.

I have been referring to principles of symmetry as giving theories a kind of rigidity. You might think that this is a drawback, that the physicist wants to develop theories that are capable of describing a wide variety of phenomena and therefore would like to discover theories that are as flexible as possible — theories that make sense under a wide variety of possible circumstances. That is true in many areas of science, but it is not true in this kind of fundamental physics. We are on the track of something universal — something that governs physical phenomena throughout the universe — something that we call the laws of nature. We do not want to discover a theory that is capable of describing all imaginable kinds of force among the particles of nature. Rather, we hope for a theory that rigidly will allow us to describe only those forces — gravitational, electroweak and strong — that actually as it happens do exist. This kind of rigidity in our physical theories is part of what we recognize as beauty.

It is not only principles of symmetry that give rigidity to our theories. On the basis of symmetry principles alone we would not be led to the electroweak theory or quantum chromodynamics, except as a special case of a much larger variety of theories with an unlimited number of adjustable constants that could be put into the theory with any values we like. The additional constraint that allowed us to pick out our simple standard model out of the variety of other more complicated theories that satisfy the same symmetry principles was the condition that infinities that arise in calculations using the theory should all cancel. (That is, the theory must be 'renormalizable.'[9]) This condition turns out to impose a high degree of simplicity on the equations of the theory and, together with the various local symmetries, goes a long way to giving a unique shape to our standard model of elementary particles.

The beauty that we find in physical theories like general relativity or the standard model is very like the beauty conferred on some works of art by the sense of inevitability that they give us — the sense that one would

---

[9]There has been a change in our thinking on this point. Today we would restrict ourselves to renormalizable interactions because it is thought that any other interactions are suppressed by inverse powers of the large energies associated with a more fundamental underlying theory. But we still use renormalizability as a guide in formulating our theories.

not want to change a note or a brush stroke or a line. But just as in our appreciation of music or painting or poetry, this sense of inevitability is a matter of taste and experience and cannot be reduced to formula.

Every other year the Lawrence Berkeley Laboratory publishes a little booklet that lists the properties of the elementary particles as known to that date. If I say that the fundamental principle governing nature is that the elementary particles have the properties listed in this booklet, then it is certainly true that the known properties of the elementary particles follow inevitably from this fundamental principle. This principle even has predictive power — every new electron or proton created in our laboratories will be found to have the mass and charge listed in the booklet. But the principle itself is so ugly that no one would feel that anything had been accomplished. Its ugliness lies in its lack of simplicity and inevitability — the booklet contains thousands of numbers, any one of which could be changed without making nonsense of the rest of the information. There is no logical formula that establishes a sharp dividing line between a beautiful explanatory theory and a mere list of data, but we know the difference when we see it — we demand a simplicity and rigidity in our principles before we are willing to take them seriously. Thus not only is our aesthetic judgment a means to the end of finding scientific explanations and judging their validity — *it is part of what we mean by an explanation.*

Other scientists sometimes poke fun at elementary particle physicists because there are now so many so-called elementary particles that we have to carry the Berkeley booklet around with us to remind us of all the particles that have been discovered. But the mere number of particles is not important. As Abdus Salam has said, it is not particles or forces with which nature is sparing, but principles. The important thing is to have a set of simple and economical principles that explain why the particles are what they are. It *is* disturbing that we do not yet have a complete theory of the sort we want. But, when we do, it will not matter very much how many kinds of particle or force it describes, as long as it does so beautifully, as an inevitable consequence of simple principles.

The kind of beauty that we find in physical theories is of a very limited sort. It is, as far as I have been able to capture it in words, the beauty of simplicity and inevitability — the beauty of perfect structure, the beauty of everything fitting together, of nothing being changeable, of logical rigidity. It is a beauty that is spare and classic, the sort we find in the Greek tragedies. But this is not the only kind of beauty that we find in the arts. A play of Shakespeare does not have this beauty, at any rate not to the extent that some of his sonnets have. Often the director of a Shakespeare play chooses to leave out whole speeches. In the Olivier film version of *Hamlet*, Hamlet never says 'Oh what a rogue and peasant slave am I!. . . .' And yet the performance works, because Shakespeare's plays are not spare, perfect structures like general relativity or *Oedipus Rex*; they are big messy compositions whose messiness mirrors the complexity of life. That is part of the beauty of his plays, a beauty that to my taste is of a higher order than the beauty of a play of Sophocles or the beauty of general relativity for that matter. Some of the greatest moments in Shakespeare are those in which he deliberately abandons the model of Greek tragedy and introduces an extraneous comic proletarian — a doorkeeper or gardener or fig seller or gravedigger — just before his major characters encounter their fates. Certainly the beauty of theoretical physics would be a very bad exemplar for the arts, but such as it is it gives us pleasure and guidance.

There is another respect in which it seems to me that theoretical physics is a bad model for the arts. Our theories are very esoteric — necessarily so, because we are forced to develop these theories using a language, the language of mathematics, that has not become part of the general equipment of the educated public. Physicists generally do not like the fact that our theories are so esoteric. On the other hand, I have occasionally heard artists talk proudly about their work being accessible only to a band of cognoscenti and justify this attitude by quoting the example of physical theories like general relativity that also can be understood only by initiates. Artists like physicists may not always be able to make themselves understood by the general public, but esotericism for its own sake is just silly.

Although we seek theories that are beautiful because of a rigidity imposed on them by simple underlying principles, creating a theory is not simply a matter of deducing it mathematically from a set of preordained principles. Our principles are often invented as we go along, sometimes precisely because they lead to the kind of rigidity we hope for. I have no doubt that one of the reasons that Einstein was so pleased with his idea about the equivalence of gravitation and inertia was that this principle led to only one fairly rigid theory of gravitation, and not to an infinite variety of possible theories of gravitation. Deducing the consequences of a given set of well-formulated physical principles can be difficult or easy, but it is the sort of thing that physicists learn to do in graduate school and that they generally enjoy doing. The creation of *new* physical principles is agony and apparently cannot be taught.

Weirdly, although the beauty of physical theories is embodied in rigid mathematical structures based on simple underlying principles, the structures that have this sort of beauty tend to survive even when the underlying principles are found to be wrong. A good example is Dirac's theory of the electron. Dirac in 1928 was trying to rework Schrödinger's version of quantum mechanics in terms of particle waves so that it would be consistent with the special theory of relativity. This effort led Dirac to the conclusions that the electron must have a certain spin, and that the universe is filled with unobservable electrons of negative energy, whose *absence* at a particular point would be seen in the laboratory as the presence of an electron with the opposite charge, that is, an antiparticle of the electron. His theory gained an enormous prestige from the 1932 discovery in cosmic rays of precisely such an antiparticle of the electron, the particle now called the positron. Dirac's theory was a key ingredient in the version of quantum electrodynamics that was developed and applied with great success in the 1930s and 1940s. But we know today that Dirac's point of view was largely wrong. The proper context for the reconciliation of quantum mechanics and special relativity is not the sort of relativistic version of Schrödinger's wave mechanics that Dirac sought, but the more general formalism known as quantum field theory, presented by Heisenberg and Pauli in 1929. In quantum field theory not only is the photon a bundle of the energy of a field, the electromagnetic field; so also the electron and positrons are bundles of the energy of the electron field, and all other elementary particles are bundles of the energy of various other fields. Almost by accident, Dirac's theory of the electron gave the same results as quantum field theory for processes involving only electrons, positrons and/or photons. But quantum field theory is more general — it can account for processes like nuclear beta decay that could not be understood along the lines of Dirac's theory.[10] There is nothing in quantum field theory that requires particles to have any particular spin. The electron does happen to have the spin that Dirac's theory required, but there are other particles with other spins and those other particles have antiparticles and this has nothing to do with the negative

---

[10] In Dirac's theory electrons are eternal; a process like the production of an electron and a positron is interpreted as the lifting of a negative-energy electron to a state of positive energy, leaving a hole in the sea of negative-energy electrons that is observed as a positron, and the annihilation of an electron and positron is interpreted as the falling of an electron into such a hole. In nuclear beta decay electrons are created *without positrons* out of the energy and the electric charge in the electron field.

energies about which Dirac speculated.[11] Yet the *mathematics* of Dirac's theory has survived as an essential part of quantum field theory; it must be taught in every graduate course in advanced quantum mechanics. The formal structure of Dirac's theory has thus survived the death of the principles of relativistic wave mechanics that Dirac followed in being led to his theory.

So the mathematical structures that physicists develop in obedience to physical principles have an odd kind of portability. They can be carried over from one conceptual environment to another and serve many different purposes, like the clever bones in your shoulders that in another animal would be the joint between the wing and the body of a bird or the flipper and body of a dolphin. We are led to these beautiful structures by physical principles, but the beauty sometimes survives when the principles themselves do not.

A possible explanation was given by Niels Bohr.[12] In speculating in 1922 about the future of his earlier theory of atomic structure, he remarked that 'mathematics has only a limited number of forms that we can adapt to Nature, and it can happen to one that he finds the right forms by formulating entirely wrong concepts.' As it happened, Bohr was right about the future of his own theory; its underlying principles have been abandoned, but we still use some of its language and methods of calculation.

It is precisely in the application of pure mathematics to physics that the effectiveness of aesthetic judgments is most amazing. It has become a commonplace that mathematicians are driven in their work by the wish to construct formalisms that are conceptually beautiful. The English mathematician G.H. Hardy explained that 'mathematical patterns like those of the painters or the poets must be beautiful. The ideas, like the colors or the words must fit together in a harmonious way. Beauty is the first test. There

---

[11] Dirac and I were at a conference in Florida in the early 1970s, and I took the occasion to ask him how he could explain the fact that there are particles (like the pi meson or the W particle) that have a spin different from the electron's, and could not have stable states of negative energy, and yet have distinct antiparticles. Dirac said that he had never thought that these particles were important.

[12] This is a recollection of Heisenberg, quoted by Valentine Telegdi and Victor Weisskopf in a review of Heisenberg's collected works in *Physics Today* (July 1991), p. 58. The same notion of the limited variety of possible mathematical forms has been expressed by the mathematician Andrew Gleason.

is no permanent place for ugly mathematics.'[13] And yet mathematical structures that confessedly are developed by mathematicians because they seek a sort of beauty are often found later to be extraordinarily valuable by the physicist.

For illustration, let us turn to the example of non-Euclidean geometry and general relativity. After Euclid, mathematicians tried for two millennia to learn whether the different assumptions underlying Euclid's geometry were logically independent of each other. If the postulates were not independent, if some of them could be deduced from the others, then the unnecessary postulates could be dropped, yielding a more economical and hence more beautiful formulation of geometry. This effort came to a head in the early years of the nineteenth century, when 'the prince of geometers' Carl Friedrich Gauss and others developed a non-Euclidean geometry for a sort of curved space[14] that satisfied all Euclid's postulates except the fifth.[15] This showed that Euclid's fifth postulate is indeed logically independent of the other postulates. The new geometry was developed in order to settle a historic question about the foundations of geometry, not at all because anyone thought it applied to the real world.

Non-Euclidean geometry was then extended by one of the greatest of all mathematicians, Georg Friedrich Bernhard Riemann, to a general theory of curved spaces of two or three or any number of dimensions. Mathematicians continued to work on Riemannian geometry because it was so beautiful, without any idea of physical applications. Its beauty was again largely the beauty of inevitability. Once you start thinking about curved spaces, you are led almost inevitably to the introduction of the mathematical quantities ('metrics,' 'affine connections,' 'curvature tensors,' and so on), that are the ingredients of Riemannian geometry. When Einstein started to

---

[13]Throughout his life Hardy boasted that his research in pure mathematics could not possibly have any practical application. But when Kerson Huang and I were working at M.I.T. on the behavior of matter at extremely high temperature, we found just the mathematical formulas we needed in Hardy's papers with Ramanujan on number theory.
[14]The principal other architects of this curved space were Janos Bolyai and Nicolai Ivanovitch Lobachevski. The work of Gauss, Bolyai and Lobachevski was important for the future of mathematics because they described this space as being not merely curved the way the surface of the earth is curved, by the way that the surface is embedded in a higher dimensional uncurved space, but in terms of its intrinsic curvature, without any reference to how it is embedded in higher dimensions.
[15]Euclid's fifth postulate in one version states that through any given point outside any given line, one and only one line can be drawn that is parallel to the given line. In the new non-Euclidean geometry of Gauss, Bolyai and Lobachevski, many such parallel lines can be drawn.

develop general relativity, he realized that one way of expressing his ideas about the symmetry that relates different frames of reference was to ascribe gravitation to the curvature of space-time. He asked a friend, Marcel Grossman, whether there existed any mathematical theory of curved spaces — not merely of curved two-dimensional surfaces in ordinary Euclidean three-dimensional space, but of curved three-dimensional spaces, or even curved four-dimensional space-times. Grossman gave Einstein the good news that there did in fact exist such a mathematical formalism, the one developed by Riemann and others, and taught him this mathematics, which Einstein then incorporated into general relativity. The mathematics was there waiting for Einstein to make use of, although I believe that Gauss and Riemann and the other differential geometers of the nineteenth century had no idea that their work would ever have any application to physical theories of gravitation.

An even stranger example is provided by the history of internal symmetry principles. In physics internal symmetry principles typically impose a kind of family structure on the menu of possible particles. The first known example of such a family was provided by the two types of particle that make up ordinary atomic nuclei, the proton and neutron. Protons and neutrons have very nearly the same mass, so, once the neutron was discovered by James Chadwick in 1932, it was natural to suppose that the strong nuclear forces (which contribute to the neutron and proton masses) should respect a simple symmetry: The equations governing these forces should preserve their form if everywhere in these equations the roles of neutrons and protons are reversed. This would tell us among other things that the strong nuclear force is the same between two neutrons as between two protons but would tell us nothing about the force between a proton and a neutron. It was therefore somewhat a surprise when experiments[16] in 1936 revealed that the nuclear force between two protons is about the same as the force between a proton and a neutron. This observation gave rise to the idea of a symmetry that goes beyond mere interchanges of protons and neutrons, a symmetry under continuous transformations that change protons and neutrons into particles that are proton-neutron mixtures, with arbitrary probabilities of being a proton or a neutron.

These symmetry transformations act on the particle label that distinguishes protons and neutrons in a way that is mathematically the same as the way that ordinary rotations in three dimensions act on the spins of

---

[16]These experiments were carried out by Merle Tuve together with N. Heydenberg and L.R. Hafstad, using a million-volt Van de Graff accelerator to fire a beam of protons into a proton-rich target like paraffin.

particles like protons or neutrons or electrons.[17] With this example in mind, until the 1960s many physicists tacitly assumed that the continuous internal symmetry transformations that leave the laws of nature unchanged had to take the form of rotations in some internal space of two, three or more dimensions, like the rotations of protons and neutrons into one another. The textbooks on the application of symmetry principles to physics then available (including the classic books of Hermann Weyl and Eugene Wigner) barely gave any hint that there were other mathematical possibilities. It was not until a host of new particles were discovered in cosmic rays and then at accelerators like the Bevatron in Berkeley in the late 1950s that a wider view of the possibilities of internal symmetries was forced on the world of theoretical physics. These particles seemed to fall into families that were more extensive than the simple neutron-proton pair of twins. For instance, the neutron and proton were found to bear a strong family likeness to six other particles known as hyperons, of the same spin and similar mass. What sort of internal symmetry could give rise to such extended kinship groups?

Around 1960 physicists studying this question began to turn for help to the literature of mathematics. It came to them as a delightful surprise that mathematicians had in a sense already catalogued all possible symmetries. The complete set of transformations that leaves anything unchanged, whether a specific object or the laws of nature, forms a mathematical structure known as a *group*,[18] and the general mathematics of symmetry

[17]For this reason, this symmetry is known as *isotopic spin symmetry*. (It was proposed in 1936 by G. Breit and E. Feenberg, and independently by B. Cassen and E.U. Condon, on the basis of the experiments of Tuve et al.) Isotopic spin symmetry is also mathematically similar to the internal symmetry that underlies the weak and electromagnetic forces in the electroweak theory, but physically quite different. One difference is that different particles are grouped into families: the proton and neutron for isotopic spin symmetry, and the left-handed electron and neutrino as well as left-handed up and down quarks for the electroweak symmetry. Also, the electroweak symmetry states the invariance of the laws of nature under transformations that can depend on position in space and time; the equations governing nuclear physics preserve their form only if we transform protons and neutrons into each other in the same way everywhere and at all times. Finally the isotopic spin symmetry is only approximate, and is understood today as a somewhat accidental consequence of the small masses of quarks in our modern theory of strong nuclear forces; the electroweak symmetry is exact and taken as a fundamental principle in the electroweak theory.

[18]If two transformations leave something unchanged then so does their 'product,' defined by performing one transformation and then the other. If a transformation leaves something unchanged, then so does its 'inverse,' the transformation that undoes the first. Also, there is always one transformation that leaves anything unchanged, the transformation that does nothing at all, known as the unit transformation because it acts like multiplication by the number one. These three properties are what make any set of operations a group.

transformations is known as *group theory*. Each group is characterized by abstract mathematical rules that do not depend on what it is that is being transformed, just as the rules of arithmetic do not depend on what it is we are adding or multiplying. The menu of the types of families that are allowed by any particular symmetry of the laws of nature is completely dictated by the mathematical structure of the symmetry group.

Those groups of transformations that act continuously, like rotations in ordinary space or the mixing of electrons and neutrinos in the electroweak theory, are called *Lie groups*, after the Norwegian mathematician Sophus Lie. The French mathematician Élie Cartan had in his 1894 thesis given a list of all of the 'simple' Lie groups, from which all others could be built up by combining their transformations.[19] In 1960 Gell-Mann and the Israeli physicist Yuval Ne'eman independently found that one of these simple Lie groups (known as $SU(3)$) was just right to impose a family structure on the crowd of elementary particles much like what had been found experimentally. Gell-Mann borrowed a term from Buddhism and called this symmetry principle the eightfold way, because the better known particles fell into families with eight members, like the neutron, proton and their six siblings. Not all families were then complete; a new particle was needed to complete a family of ten particles that are similar to neutrons and protons and hyperons but have three times higher spin. It was one of the great successes of the new $SU(3)$ symmetry that this predicted particle was subsequently discovered in 1964 at Brookhaven[20] and turned out to have the mass estimated by Gell-Mann.

Yet this group theory that turned out to be so relevant to physics had been invented by mathematicians for reasons that were strictly internal to mathematics. Group theory was initiated in the early nineteenth century by Evariste Galois, in his proof that there are no general formulas for the solution of certain algebraic equations (equations that involve fifth or higher powers of the unknown quantity).[21] Neither Galois nor Lie nor Cartan had any idea of the sort of application that group theory would have in physics.

It is very strange that mathematicians are led by their sense of mathematical beauty to develop formal structures that physicists only later

---

[19]Briefly, there are three infinite categories of simple Lie groups: the familiar rotation groups in two, three or more dimensions, and two other categories of transformations somewhat like rotations, known as unitary and symplectic transformations. In addition there are just five 'exceptional' Lie groups that do not belong to any of these categories.
[20]By a group headed by N. Samios.
[21]The group in question in Galois's work was the set of permutations of the solutions of the equation.

find useful, even where the mathematician had no such goal in mind. A well-known essay by the physicist Eugene Wigner[22] refers to this phenomenon as 'The Unreasonable Effectiveness of Mathematics.' Physicists generally find the ability of mathematicians to anticipate the mathematics needed in the theories of physicists quite uncanny. It is as if Neil Armstrong in 1969 when he first set foot on the surface of the moon had found in the lunar dust the footsteps of Jules Verne.

Where then *does* a physicist get a sense of beauty that helps not only in discovering theories of the real world, but even in judging the validity of physical theories, sometimes in the teeth of contrary experimental evidence? And how does a mathematician's sense of beauty lead to structures that are valuable decades or centuries later to physicists, even though the mathematician may have no interest in physical applications?

There seem to me to be three plausible explanations, two of them applicable throughout much of science and the third limited to the most fundamental areas of physics. The first explanation is that the universe itself acts on us as a random, inefficient, and yet in the long run effective, teaching machine. Just as through an infinite series of accidental events, atoms of carbon and nitrogen and oxygen and hydrogen joined together to form primitive forms of life that later evolved into protozoa and fishes and people, in the same way our way of looking at the universe has gradually evolved through a natural selection of ideas. Through countless false starts, we have gotten it beaten into us that nature is a certain way, and we have grown to look at that way that nature is as beautiful.

I suppose this would be everyone's explanation of why the horse trainer's sense of beauty helps when it does help in judging which horse can win races. The racehorse trainer has been at the track for many years — has experienced many horses winning or losing — and has come to associate, without being able to express it explicitly, certain visual cues with the expectation of a winning horse.

One of the things that makes the history of science so endlessly fascinating is to follow the slow education of our species in the sort of beauty to expect in nature. I once went back to the original literature of the 1930s on the earliest internal symmetry principle in nuclear physics, the symmetry that I mentioned earlier between neutrons and protons, to try to find the one research article that first presented this symmetry principle the way

---

[22]E.P. Wigner, 'The Unreasonable Effectiveness of Mathematics,' *Communications in Pure and Applied Mathematics*, 13 (1960), pp. 1–14.

it would be presented today, as a fundamental fact about nuclear physics that stands on its own, independent of any detailed theory of nuclear forces. I could find no such article. It seems that in the 1930s it was simply not good form to write papers based on symmetry principles. What was good form was to write papers about nuclear forces. If the forces turned out to have a certain symmetry, so much the better, for, if you knew the proton-neutron force, you did not have to guess the proton-proton force. But the symmetry principle itself was not regarded, as far as I can tell, as a feature that would legitimize a theory — that would make the theory beautiful. Symmetry principles were regarded as mathematical tricks; the real business of physicists was to work out the dynamical details of the forces we observe.

We feel differently today. If experimenters were to discover some new particles that formed families of some sort or other like the neutron-proton doublet then the mail would instantly be filled with hundreds of preprints of theoretical articles speculating about the sort of symmetry that underlies this family structure, and, if a new kind of force were discovered, we would all start speculating about the symmetry that dictates the existence of that force. Evidently we have been changed by the universe acting as a teaching machine and imposing on us a sense of beauty with which our species was not born.

Even mathematicians live in the real universe, and respond to its lessons. Euclid's geometry was taught to schoolchildren for two millennia as a nearly perfect example of abstract deductive reasoning, but we learned in this century from general relativity that Euclidean geometry works as well as it does only because the gravitational field on the surface of the earth is rather weak, so that the space in which we live has no noticeable curvature. In formulating his postulates Euclid was in fact acting as a physicist, using his experience of life in the weak gravitational fields of Hellenistic Alexandria to make a theory of uncurved space. He did not know how limited and contingent his geometry was. Indeed, it is only relatively recently that we have learned to make a distinction between pure mathematics and the science to which it is applied. The Lucasian Chair at Cambridge that was held by Newton and Dirac was (and still is) officially a professorship in mathematics, not physics. It was not until the development of a rigorous and abstract mathematical style by Augustin-Louis Cauchy and others in the early nineteenth century that mathematicians took as an ideal that their work should be independent of experience and common sense.[23]

---

[23]J.L. Richards, 'Rigor and Clarity: Foundations of Mathematics in France and England, 1800–1840,' *Science in Context*, 4 (1991), p. 297.

The second of the reasons why we expect successful scientific theories to be beautiful is simply that scientists tend to choose problems that are likely to have beautiful solutions. The same may even apply to our friend the racehorse trainer. He trains horses to win races; he has learned to recognize which horses are likely to win and he calls these horses beautiful; but, if you take him aside and promise not to repeat what he says, he may confess to you that the reason he went into the business of training horses to win races in the first place was because the horses that he trains are such beautiful animals.

A good example in physics is provided by the phenomenon of smooth phase transitions,[24] like the spontaneous disappearance of magnetism when an iron permanent magnet is heated above a temperature of 770°C, the temperature known as the Curie point. Because this is a smooth transition, the magnetization of a piece of iron goes to zero gradually as the temperature approaches the Curie point. The surprising thing about such phase transitions is the *way* that the magnetization goes to zero. Estimates of various energies in a magnet had led physicists to expect that, when the temperature is only slightly below the Curie point, the magnetization would be simply proportional to the square root of the difference between the Curie point and the temperature. Instead it was observed experimentally that the magnetization is proportional to the 0.37 power of this difference. That is, the dependence of the magnetization on the temperature is somewhere between being proportional to the square root (the 0.5 power) and the cube root (the 0.33 power) of the difference between the Curie point and the temperature.

Powers like this 0.37 are called *critical exponents*, sometimes with the adjective 'nonclassical' or 'anomalous,' because they are not what had been expected. Other quantities were observed to behave in similar ways in this and other phase transitions, in some cases with precisely the same critical exponents. This is not an intrinsically glamorous phenomenon, like black holes or the expansion of the universe. Nevertheless, some of the brightest theoretical physicists in the world worked on the problem of the critical

---

[24]What I call 'smooth' phase transitions are also often called 'second-order phase transitions.' This is to distinguish them from 'first-order phase transitions,' like the boiling of water at 100°C or the melting of ice at 0°C, in which the properties of the material change discontinuously. It takes a certain amount of energy (the so-called latent heat) to convert ice at 0°C to liquid water at the same temperature, or liquid water at 100°C to water vapor at the same temperature, but it takes no extra energy to wipe out the magnetism of a piece of iron when the temperature is just at the Curie point.

exponents, until the problem was eventually solved in 1972 by Kenneth Wilson and Michael Fisher, both then at Cornell. Yet it might have been thought that the precise calculation of the Curie point itself was a problem of greater practical importance. Why should leaders of condensed matter theory give the problem of the critical exponents so much greater priority?

I think that the problem of critical exponents attracted so much attention because physicists judged that it would be likely to have a beautiful solution. The clues that suggested that the solution would be beautiful were above all the universality of the phenomenon, the fact that the same critical exponents would crop up in very different problems, and also the fact that physicists have become used to finding that the most essential properties of physical phenomena are often expressed in terms of laws that relate physical quantities to powers of other quantities, such as the inverse-square law of gravitation. As it turned out, the theory of critical exponents has a simplicity and inevitability that makes it one of the most beautiful in all of physics. In contrast, the problem of calculating the precise temperatures of phase transitions is a messy one, whose solution involves complicated details of the iron or other substance that undergoes the phase transition, and for this reason it is studied either because of its practical importance or in want of anything better to do.

In some cases the initial hopes of scientists for a beautiful theory have turned out to be misplaced. A good example is provided by the genetic code. Francis Crick describes in his autobiography[25] how after the discovery of the double helix structure of DNA by himself and James Watson, the attention of molecular biologists turned to breaking the code by which the cell interprets the sequence of chemical units on the two helices of DNA as a recipe for building suitable protein molecules. It was known that proteins are built up out of chains of amino acids, that there are only twenty amino acids that are important in virtually all plants and animals, that the information for selecting each successive amino acid in a protein molecule is carried by the choices of three successive pairs of chemical units called bases, of which there are only four different kinds. So the genetic code interprets three successive choices each out of four possible base pairs (like three cards chosen in order from a deck of cards that show only the four suits but no numbers or faces) to dictate each choice of one out of twenty possible amino acids to be added to the protein. Molecular biologists

[25]F. Crick, *What Mad Pursuit: A Personal View of Scientific Discovery*, New York: Basic Books, 1988.

invented all sorts of elegant principles that might govern this code — for instance, that no information in the choice of three base pairs would be wasted, and that any information not needed to specify an amino acid would be used for error detection, like the extra bits that are sent between computers to check the accuracy of the transmission. The answer found in the early 1960s turned out to be very different. The genetic code is pretty much of a mess; some amino acids are called for by more than one triplet of base pairs, and some triplets produce nothing at all.[26] The genetic code is not as bad as a randomly chosen code, which suggests that it has been somewhat improved by evolution, but any communications engineer could design a better code. The reason of course is that the genetic code was *not* designed; it developed through a series of accidents at the beginning of life on earth and has been inherited in more or less this form by all subsequent organisms. Of course the genetic code is so important to us that we study it whether it is beautiful or not, but it is a little disappointing that it did not turn out to be beautiful.

Sometimes when our sense of beauty lets us down, it is because we have overestimated the fundamental character of what we are trying to explain. A famous example is the work of the young Johannes Kepler on the sizes of the orbits of planets.

Kepler was aware of one of the most beautiful conclusions of Greek mathematics, concerning what are called the Platonic solids. These are three-dimensional objects with plane boundaries, for which every vertex and every face and every line are just like every other vertex, face and line. An obvious example is the cube. The Greeks discovered that there are all together only five of these Platonic solids: the cube, the triangular pyramid, the twelve-sided dodecahedron, the eight-sided octahedron, and the twenty-sided icosahedron. (They are called Platonic solids because Plato in the *Timaeus* proposed a one-to-one correspondence between them and the supposed five elements, a view subsequently attacked by Aristotle.) The Platonic solids furnish a prime example of mathematical beauty; this discovery has the same sort of beauty as the Cartan catalog of all possible continuous symmetry principles.

Kepler in his *Mysterium cosmographicum* proposed that the existence of just five Platonic solids explained why there were just six planets: Mercury, Venus, Earth, Mars, Jupiter, and Saturn. (Uranus, Neptune, and Pluto were not discovered until later.) Each one of the Platonic solids was conceived

---

[26]Strictly speaking, the otherwise meaningless triplets do carry the message 'end chain.'

to just fit between the spheres carrying two of the planets. With the solids nested between the spheres in the right order, the ratios of the radii of all the planetary orbits could be predicted.

To a scientist today it may seem a scandal that one of the founders of modern science should invent such a fanciful model of the solar system. This is not only because Kepler's scheme did not fit observations of the solar system (though it did not), but much more because we know that this is not the sort of speculation that is appropriate to the solar system. But Kepler was not a fool. The kind of speculative reasoning he applied to the solar system is very similar to the sort of theorizing that elementary particle physicists do today; we do not associate anything with the Platonic solids, but we do believe for instance in a correspondence between different possible kinds of force and different members of the Cartan catalog of all possible symmetries. Where Kepler went wrong was not in using this sort of guesswork, but in supposing (as most philosophers before him had supposed) that the planets are important.

Of course, the planets are important in some ways. We live on one of them. But their existence is not incorporated into the laws of nature at any fundamental level. We now understand that the planets and their orbits are the results of a sequence of historical accidents and that, although physical theory can tell us which orbits are stable and which would be chaotic, there is no reason to expect any relations among the sizes of their orbits that would be mathematically simple and beautiful.

It is when we study truly fundamental problems that we expect to find beautiful answers. We believe that, if we ask why the world is the way it is, and then ask why that answer is the way it is, at the end of this chain of explanations we shall find a few simple principles of compelling beauty. We think this in part because our historical experience teaches us that as we look beneath the surface of things, we find more and more beauty. Plato and the neo-Platonists taught that the beauty we see in nature is a reflection of the beauty of the ultimate, the *nous*. For us, too, the beauty of present theories is an anticipation, a premonition, of the beauty of the final theory. And in any case, we would not accept any theory as final unless it were beautiful.

Although we do not yet have a sure sense of where in our work we should rely on our sense of beauty, still in elementary particle physics aesthetic judgments seem to be working increasingly well. I take this as evidence that we are moving in the right direction, and perhaps not so far from our goal.

Prof. Steven Weinberg delivered the Albert Einstein Memorial Lecture in 1984.

Harmless Energy from Nuclei

Carlo Rubbia

1. Energy Is Necessary

The interest of politicians, businessmen, technologists, scientists and the public at large is focused today on the problem of energy. All will agree to the assertion that 'energy is necessary' for the future of humankind, but many tend to paraphrase it by saying that 'energy is a necessary evil.' Necessary it surely is, but an analysis of the motivations for regarding energy as 'evil' reveals some Freudian undertones. After centuries of a passionate technological endeavor that has become deeply engraved upon our conception of the world, this rejection of technology as a solution to rising environmental concerns, perceived as a Faustian deal, is a curious phenomenon, to say the least. To be sure, the concerns associated with energy production are serious, and the inevitable growth of energy consumption, pushed by the sheer momentum of the system and the very human expectations of the poor, may indeed add enough yeast to make the ensuing problems leaven beyond control. However, it is our responsibility as rational people to go beyond the arguments for impending doom and to moderate the rhetoric. Emotions tend to blur the image. As in the areas of famine, disease and so on, here, too, science and technology should be trusted. Indeed, we may reasonable expect that, combined, they will be able to solve the energy problem as well, in full accord with the economic, dynamic and technical constraints with which a working system has to comply.

That energy supply has been a major element of our civilization is evidenced in Figure 1, where approximate energy use per capita since the emergence of humankind is shown as a function of time. Energy required for food gathering was supplemented by that for household use (initially heating), organized agriculture, industry and transportation. The

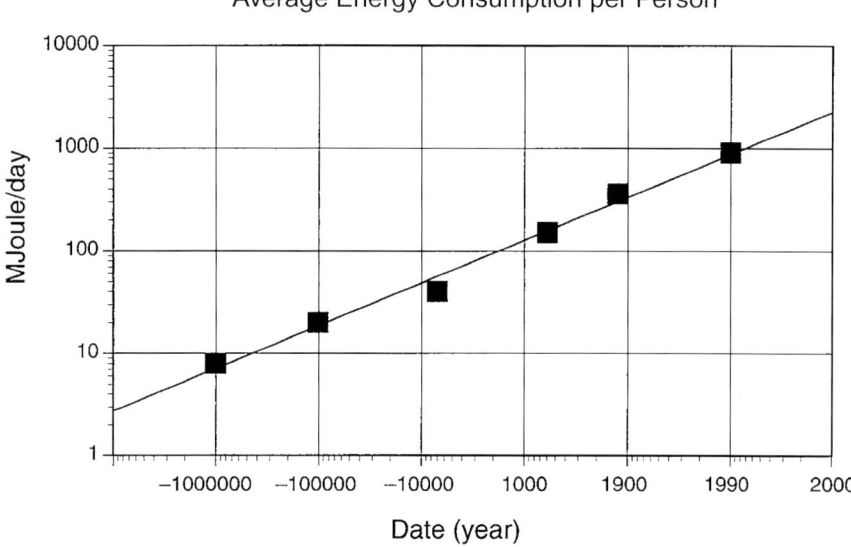

Figure 1. Energy vs. time for the technologically advanced segment of human civilization. Approximate per capita consumption of energy as a function of time [R.A. Knief, 1992]. Energy for food gathering has been supplemented sequentially by that for household use (initially heating), organized agriculture, industry and transportation.

consumption of hay by working horses[1] — the equivalent of diesel fuel for trucks and tractors today — is included. We see that total energy consumption for the most technologically advanced sector of humanity has grown about a hundred-fold from the beginning of history, to reach today's level of about 0.9 GJ/day/person. This corresponds to an effective 32 kg of coal/day/person, or a continuous, averaged supply of 10.4 kWatt/person. The direct total human-generated energy production of the planet, mostly derived from fossil fuels, corresponds to an average total daily power production in excess of 10 teraWatts. By comparison, the geological heat emanating from the earth's crust on account of natural uranium and thorium decay is about 16 tWatt. The presence of humans on the planet has roughly doubled its energy generation.

---

[1] As late as 1899, about two thirds of the mechanical energy consumed in the USA still came from horses.

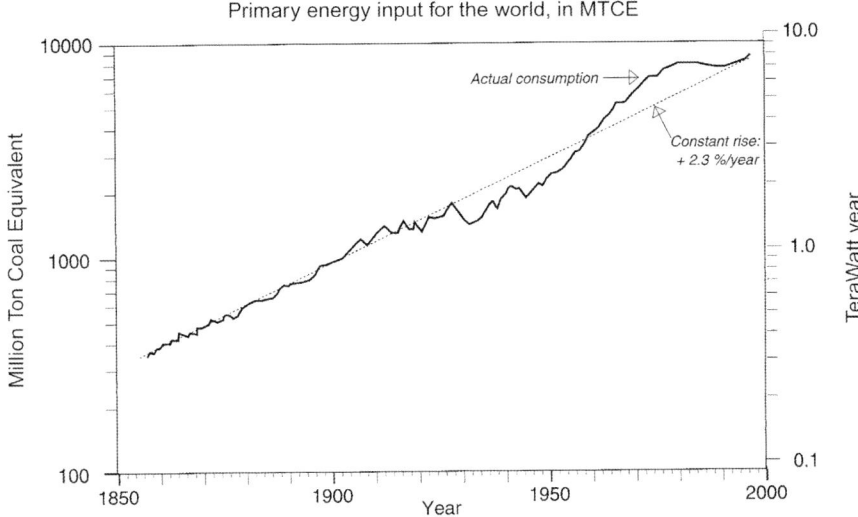

Figure 2. Energy consumption as a function of time.

The energy consumption of the planet has steadily increased at the rate of 2%/year over the last 150 years (Figure 2). There is no doubt that the world's consumption will continue to grow in the future, since the world's population is steadily growing, and billions of people in the developing countries are striving for a better life. The present enormous disparity in energy consumption — Swedes use some 15,000 kWatt hours of electricity/person/year, Tanzanians only 100 kWatt h/p/y — will tend to diminish. There is also no doubt that energy will have to be produced and used in a more efficient way. However, this is a necessary but not sufficient condition for the stabilization of energy consumption. We will undoubtedly get more mileage out of a liter of petrol, but there will be more cars; light bulbs will be more efficient, but there will be more light bulbs; and so on. Energy consumption will be more efficient, but will continue to increase. What we call 'energy intensity,' that is, kWatt hours per dollar earned, is known to be roughly constant, slowly varying with social conditions and time. The world's economic forecast is for a GNP growth of about 2%/year. It is not an accident that this is also roughly the expected planet-wide growth in energy consumption.

Such a large consumption raises obvious questions regarding the longevity of (fossil) resources. This is displayed in Figure 3, which shows

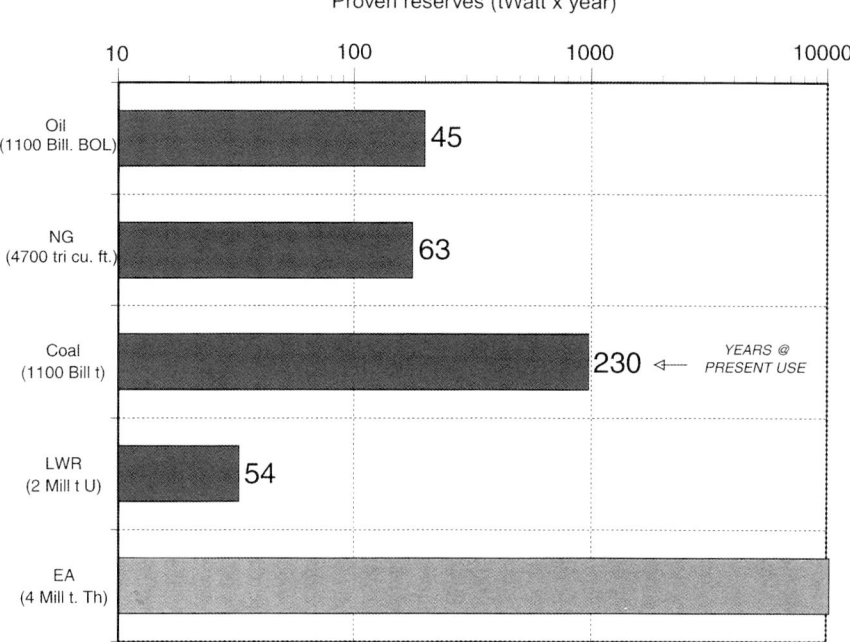

Figure 3. Proven resources and duration of energy sources.

the expected consumption in tWatt × year for each major energy source and its expected duration at the present level of consumption. There is no doubt that in order to sustain the pace of our civilization's growth, some new massive energy sources will be needed in the long run. Though the renewables (solar energy, wind, etc.) have a very important role to play in this regard, they alone may not be enough to sustain all future needs.

The fundamental scientific contribution of Einstein, namely, his realization of the existence of the potentially immense energy to be harnessed inside nuclei, becomes one of his fundamental contributions to the peaceful future of humankind. But energy from nuclei does not necessarily mean nuclear power on the present model. Used as it is today, as we shall see, uranium is no more abundant a resource than gas or oil. New reactions and new devices that use the fuel differently are necessary. This will be the main subject of my presentation. But let me first briefly address the question of climatic changes.

2. The Greenhouse Issue

There is no doubt that in the medium-long term, the issue of the availability of fossil fuels — with the possible exception of coal, of which there is an ample supply at least for a few more centuries — will be a matter of concern. However, a premature deadline may have to be set for the utilization of fossils on account of environmental considerations. Humanity's main alternatives with regard to the greenhouse effect are to fight it or to accept it. To be sure, a massive greenhouse effect will not necessarily make tomorrow's world worse for everybody: Siberia and northern Canada may have much to gain, while others, like the Mediterranean countries, stand to be afflicted with an increased incidence of tropical illnesses, desertification, drought, deforestation, and so on. Clearly, the greenhouse effect will make tomorrow's world very different, and, what is more worrisome, substantially unpredictable.

It is evident that, for the first time in history, human activities have begun to modify the global conditions of the planet. We begin to realize that environmental effects make the *price* that the community ultimately must pay for a barrel of oil significantly higher than the *cost* charged by the producers. These lately widespread preoccupations have recently been formally, politically recognized (I could say rubber-stamped) in the protocol of the Kyoto Conference, which has introduced a new dimension to the energy problem. This, in turn, has indicated the existence of a potentially much closer limit to fossil utilization than sheer natural supply.

Though no one disputes the global warming of the planet and the rising CO_2 content in the atmosphere, the phenomenology is very complex, and the relationship of cause to effect is still somewhat controversial. Three major factors cited in this regard are: (1) the very large amount of CO_2 exchanged between the sea and the atmosphere, some 30 times greater than human-generated emissions; (2) the presence and role of other greenhouse gases, such as methane and chloro-fluoro-compounds; and (3) the intrinsic instability of the planet's climate, which was subject to large variations even prior to the advent of our technological era. However, a very intriguing correlation is evidenced by the so-called Vostock Carrot, a sample of deep ice from the Antarctic studied by a French-Russian team (Figure 4). The average temperature of the planet over a period of 160,000 years was gauged by analyzing the isotopic composition of the captured atmospheric oxygen in the layers of ice. When the mean temperature was compared with the amount of CO_2 dissolved in the same layers, the correlation was staggering.

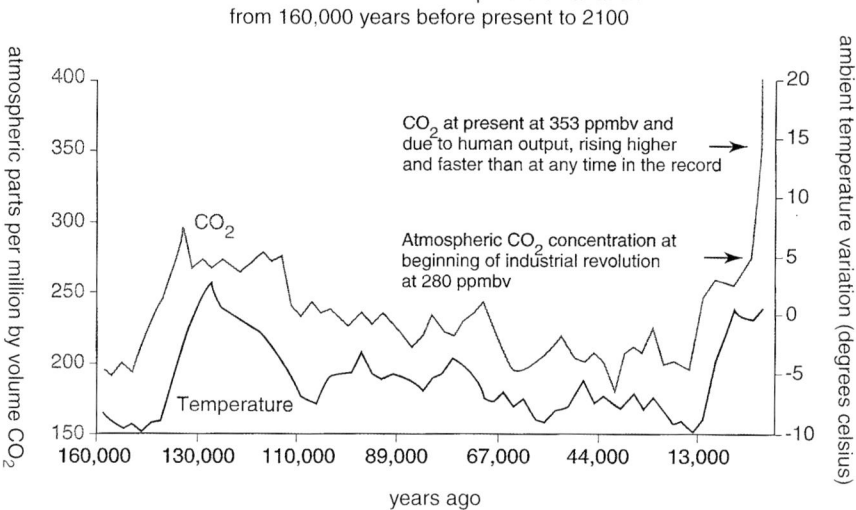

Figure 4. The Vostock Carrot: Correlation between the mean temperature of the planet and CO_2 as evidenced over the last 160,000 years.

In all events, the most elementary prudence suggests that CO_2 emissions due to human activity, which in 1990 amounted to some 15 Gtons/year, should progressively and significantly be curbed. The Kyoto Agreement mandates that the most technologically advanced countries reduce their emissions by some 5–10% over the next fifteen years, relative to the present level. However, since energy consumption is expected to continue rising at the rate of some 1.5–2.0%/year, this actually means a reduction of about one third in the projected uncurbed emissions. This is not an easy problem. Since fifteen years are not a long time in this field, both substantial policy changes and some quick-fix action are required. Moreover, if the climatic changes become more evident in the years to come, there may be additional political pressure toward considering the Kyoto Agreement as no more than a first step in the program.

Curbing CO_2 emissions does not necessarily mean a total ban on fossil fuels. Indeed, new methods are emerging for the safe disposal of CO_2 without discharging it into the atmosphere:

(1) Large amounts of CO_2 can be stored in liquid form deep underwater (>3 km). CO_2 is liquid at about 73 atmospheres (30°C), that is, at

about 700 meters ocean depth. However, this method of disposal raises many environmental concerns of its own (e.g., the impact of PH changes on ocean life and ecosystems).

(2) Depleted oil and natural gas fields can absorb CO_2 if it can be pumped back underground. For example, a European study (the Joule II Project) has shown that:

- Underground disposal is a perfectly feasible method of disposing of very large quantities of CO_2.
- There is space available in the EU and Norway to store approximately 800 Gton CO_2 without exceeding the original reservoir pressure. This is adequate to store up to 250 years of total CO_2 emissions from OECD-Europe at the 1990 rate of 3.2 ton/year.
- All the necessary technological steps are commercially proven and thus could be implemented today.
- The study of large, naturally occurring CO_2 accumulations indicates that CO_2 can be retained in underground reservoirs for millions of years.

(3) Underground storage of CO_2 can be performed in a variety of soil configurations, thus enhancing the generality of the method. For instance:

- Large aquifer(ous) beds could absorb somewhere between 50 Gton CO_2 and 14,000 Gton CO_2, depending on the conditions (structural traps) under which CO_2 may either dissolve[2] or be trapped. However, because the area to be occupied by the CO_2 will be large, uncertainties about reservoir integrity will be large and controllability of the site will be small. Adequate storage security may be a problem.
- CO_2 injection and sequestration could enable enhanced recovery of deep coal-bed methane (CMB, i.e., CH_4). Large amounts of CH_4, equivalent to 30–120% of conventional natural gas resources in some areas (China has twice as much CMB as the USA has natural gas), are contained in coal beds. CO_2 is twice as adsorbent as CH_4, and so, as pressurized CO_2 moves in the reservoir, it displaces and compresses CH_4. Very little CO_2 shows up until most of the CH_4 has been recovered.

---

[2]CO_2 saturated water is denser than pure water, and so will not rise by virtue of buoyancy.

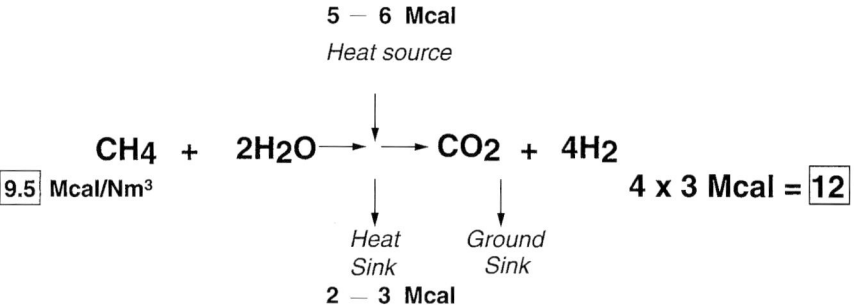

Figure 5. Gross energy balance of the steam reforming process.

These methods are feasible, but, inevitably, they will considerably increase the cost of the associated energy. For example, in the case of the underground disposal of CO_2 from fossil-fuelled power stations, the CO_2 must be separated from the flue gas before disposing of it. This will be less costly if the natural gas or coal is transformed into hydrogen by means of the well-known (endothermic) steam-reforming reaction:

$$CH_4 + 2H_2O \rightarrow CO_2 + 4H_2$$

CO_2 can be easily separated from hydrogen, with the help of a molecular sieve[3] or by other simple methods, and injected underground. The H_2 thus recovered is an excellent, clean energy carrier that permits the use of higher-efficiency electricity conversions and of fuel cells. The gross energy balance is shown in Figure 5. The energy content of the produced H_2 is larger than that of the CH_4 feed stock (+26%).

However, the method requires an external energy source which must produce about 50% of the delivered energy in the form of moderate to high-temperature heat. If some of the produced hydrogen is used to heat up the plant, the overall efficiency, relative to the CH_4 feed stock, is only about 60% (6/9.5). Taking into account the cost of the manipulations, the cost of the ultimate energy would be almost double that of conventionally produced fossil energy (about \$6/GJ for H_2, as opposed to \$3/GJ for natural gas).

---

[3] The hydrogen molecule is a tiny object of few Ås in size. The CO_2 molecule is several dozen times larger. The porous structure of a molecular sieve can easily let through the H_2 while retaining the CO_2.

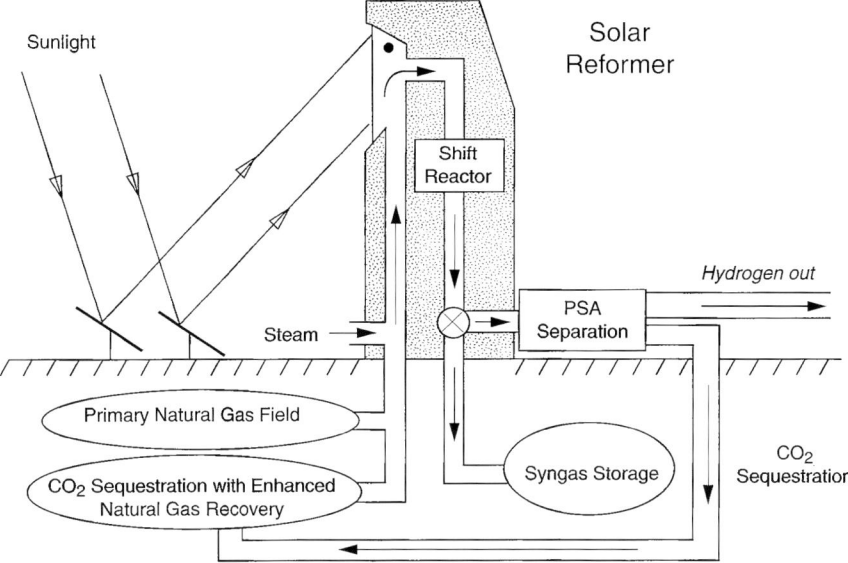

Figure 6. Diagram in principle of solar-assisted CO_2 recovery and CH_4 transformation into H_2.

In order to increase the energy efficiency and reduce costs, one could use some alternative, emissionless source of energy to produce the reforming, of which about 50% will end up in the energy content of the H_2 produced by the process. Two examples are:

(1) *High-temperature ($\geq 600°C$) solar thermal energy.* This method (Figure 6) is attractive, since about half of the solar energy and all the energy of the natural gas emerge in the form of hydrogen, while the CO_2 is separated out and eventually disposed of underground. The hydrogen thus produced is 22% solar-derived and 78% CH_4-derived.[4]

(2) *A simple and safe nuclear heat generator (EA)* of the type described later on. Again, the natural gas will double the energy yield of the nuclear source and provide the chemistry needed for the production of hydrogen, but with no CO_2 emission into the atmosphere.

---

[4]This combined method seems far more attractive than the classical methods of producing H_2 by electrolysis or by direct water decomposition at a very high temperature (1800°C), both of which are plagued by low efficiency.

There is no doubt that the need to curb CO_2 emissions will lead to a substantial increase in the cost of energy and spur the development of systems combined with CO_2 separation and disposal. These methods will, in turn, encourage the use of hydrogen as a clean energy carrier and a substitute for natural gas, as well as the use of high-efficiency fuel cells. An important asset in this substitution is the basic interchangeability of natural gas and hydrogen, in the sense that a large majority of the existing installations could be retro-fitted from one to the other.[5]

It is probable that recuperation schemes will prevail over new methods at least during a transitional period, determined by the remaining useful lifetime of the existing large installations, in which huge investments have been made (about 4% of the collective planetary GNP, or about US$ 1 trillion,[6] is spent annually on expanding existing energy systems and technologies; over an average lifetime of 20 years for these installations, they represent, collectively, an investment of some US$ 20 trillion, equivalent to the planetary GNP in 1990). The resultant complexity of these combined methods should not be underestimated. In my view, the principle that *the cheapest energy is the best energy* ultimately will prevail. Under both market and environmental pressures, all kinds of zero or low-emission energy sources will be developed — a bonanza for new entrepreneurial ideas. The remainder of my lecture will be devoted to the role that a renovated nuclear energy source may play in this new situation.

3. Pros and Cons of Present-Day Nuclear Energy

When nuclear energy was first developed in the 1960s, it was greeted with the greatest enthusiasm (we recall, for instance, the 1959 international, UN-sponsored 'Atoms for Peace' program in Geneva). It promised an unlimited, cheap and abundant source of energy for the future of humankind. In the course of the years this enthusiasm has gradually disappeared, and today nuclear power is perceived by many as 'evil.' Under the pressure of popular concern, a huge number of regulatory constraints have eroded the price margin of nuclear energy. Today it no longer seems to be 'the cheapest energy' compared with fossil fuels, particularly natural gas and coal. At least in the

---

[5]The old-fashioned 'town-gas' which is progressively being replaced by natural gas contained as much as 50% hydrogen. A pipeline designed for natural gas can transport about 70% as much energy in the form of H_2 without requiring any modification. However, the H_2 may produce metal embrittlement.
[6]This is also the full budget of the US Government.

developed countries, nuclear power also evidently has almost completely filled its potential market niche; in good company with steel, housing and cars, the number of nuclear installations seems to have reached some kind of saturation point. This is not entirely the case in the developing countries (such as China), where some new installations are nuclear. The problem will present itself again in some 15–20 years, when present installations reach the end of their practical life span.

The a priori predicted features of nuclear energy, compared with fossil fuels, are (1) potentially zero emissions and (2) an extremely parsimonious use of the fuel. For instance, one ton of uranium — provided it could be completely fissioned — could produce the equivalent in energy of fourteen million barrels of oil (BOL), or three million tons of coal (TEC). There is therefore a potential gain in the power yield of about 3×10^6 with respect to chemical energy. The present planetary demand for energy (10 tWatt) could ideally be entirely met with about 3,900 tons of fissile material per year.[7] If fission is replaced with fusion (D + T), the primary consumption of natural lithium (from which the unstable T is bred) required to meet the demand would be a mere 16,000 tons.

Unfortunately, nuclear power technology as presently constituted, essentially based on light water reactors (LWR) operated mostly on a basis of enriched uranium and thermal neutrons, is very far from such an idealized expectation. Following Héfele (see Table 1), we consider the practical effort required to run a LWR of 1 $GW_{electric}$ = 3.3 $GW_{thermal}$ for 30 years, again referred to the 'natural' fuel demand. Only the ^{235}U, comprising 0.71% of natural uranium, is thermally fissile, and about 60% of this is extracted by enrichment. Thus, only about 0.4% of the uranium is potentially useful. Taking into account the fraction of unburned fuel left in the spent fuel and the secondary breeding of plutonium, we conclude that for a standard (33 gWatt × day/t) open cycle, only $\eta = 1/200$ of the original uranium is actually fissioned. Uranium concentration in conventional ores is modest, typically 2,000 p.p.m. These more favorable ores eventually will be exhausted, and for future considerations a lower enrichment level, on the order of, say, 70 p.p.m., is probably more realistic. Referred to ores, the 2,000 p.p.m. case corresponds to an energy yield of $(3 \times 10^6) \times (2 \times 10^{-3})/200 = 30$ times that of plain coal; it drops to $(3 \times 10^6) \times (7 \times 10^{-5})/200 = 1$ for the lower-yield (70 p.p.m.) ores.

---

[7]The present production of uranium is about 100,000 tons/y.

Table 1.   Logistic effort in energy production.

Energy source: 1 GWe × 30 years 6.1 TWh	Land km^2	Mining work man-years	Ores handling tons
Ordinary reactors (LWR): High-content ores (2000 p.p.m)	3	1,500	$4.50 \times 10^{7[1]}$
Ordinary reactors (LWR): Low-content ores (70 p.p.m)	33	9,000	$3.60 \times 10^{8[1]}$
Coal-fired plant	10–20	15,000	$3.21 \times 10^{8[2]}$
Energy amplifier (EA): thorium ores at 4%	7.5×10^{-4}	0.375	$1.12 \times 10^{4[1]}$
Ratio Coal/EA	$(1.3–2.6) \times 10^4$	4.00×10^4	2.77×10^4

[1]Overburden factor, averaged: 15 m^3/ton
[2]Overburden factor, averaged: 3 m^3/ton

Table 1 makes a more complete estimate, taking into account the differences in overburden, etc.

The conclusion is that most of the 'magic' factor of 3×10^6 for nuclear energy is, as of today, almost wiped out. One could naively ask what is the advantage of using uranium, whose extraction is difficult and dangerous, among other things, because of radon emission (about 75% of the radiation passed to the public comes from mining), when — with a comparable mining effort — one could extract directly ignitable, shallow coal and burn it! Of course other factors enter into the game, such as the cost of transportation for coal and the fuel preparation for uranium, but they do not alter these basic 'inefficiency' considerations. That is why, notwithstanding the tremendous potential of nuclear energy, uranium used in this way will yield no more energy for future use than oil. In Table 1, anticipating the following pages, we also display the much more favorable situation of the so-called Energy Amplifier, an accelerator-driven device that works by burning natural thorium directly. One can see that alternative methods are indeed capable of restoring a substantial, final factor of 2.8×10^4 with respect to coal burning.

There are other important arguments that work to the detriment of a purely LWR-based nuclear energy option — especially if it has to be generalized — and need to be overcome by novel technologies:

(1) A significant amount of long-lived isotopes (gases, etc.) are released into the environment, and more if 'reprocessing' is used in order to

improve fuel efficiency. The total dose given so far to the public during normal operation (6% of the world's energy production) amounts to 2×10^5 Sievert × person.[8] This is tiny when compared to that of atmospheric nuclear test explosions (10^7 Sievert × person), the general natural background or medical needs, but it is not small in absolute number.[9] For the same generated power, these emissions (200 Sievert × person/year/gWatt$_e$) represent about ten times the radiotoxic elements emitted by coal fumes, which also contain tiny fractions of radioactive impurities.

(2) Accidents — mainly the criticality accident at Chernobyl and the core melt-down event at Three Mile Island — have almost doubled this dose. New devices should exhibit no criticality and provide deterministic (not probabilistic) insurance against accidental melt-down — for example, by using a spontaneously self-cooled configuration.

(3) The problem of long lived radioactive waste: Existing nuclear power plants annually produce about 12,000 tons of highly radioactive spent fuel, of which about 1% (120 tons) is plutonium. The radiotoxicity of this mass of material reaches the level of the initial uranium ores only after about one million years (Figure 7). It is a fortunate circumstance, to be addressed by innovative technologies, that the most offending elements (trans-uranic nuclei [TRU] and some long-lived nuclei like ^{129}I and ^{99}Tc), which carry the bulk of the long-lived radiotoxicity (99.995% or $1–2 \times 10^{-5}$ after 1,000 years), are only a few percent of the spent fuel, while the rest are either (1) elements with medium (≤ 30 years) or short-lived radioactivity, which can be left to decay naturally, or (2) stable elements, including the bulk of unburned uranium (94% of the metal mass), which can be recycled in the environment, eventually for further use. From this perspective, the TRUs should not be looked upon simply as 'waste,' provided they can be burned completely by means of some new fission or fusion device, operated with sufficiently fast neutrons to make all these elements fissionable. In this way, an additional 30% of the originally LWR-produced energy could be collected in the incineration

[8]This number includes both local and global radiation fallouts, and it is truncated after a period of 10,000 years for long-lived radio-nuclides.

[9]Exposure to 3–4 Sieverts daily doubles a person's cancer probability, roughly equivalent to two packages of cigarettes/day. The radiation harm from a 1 gWatt electric power station corresponds to the smoking of 400 packages of cigarettes/day by the population. However this figure is obtained neglecting the problem of waste disposal and the complex 'enrichment' biological processes which may 'channel' radioactivity into the biological cycle, as is known to be the case for ^{129}I and ^{99}Tc.

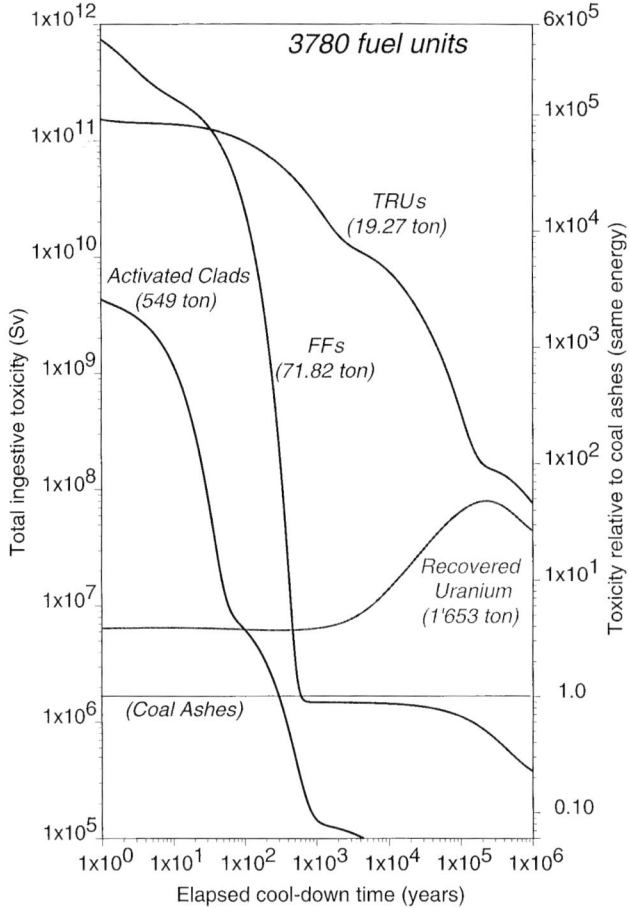

Figure 7. Energy vs. time for the technologically advanced segment of human civilization.

process. In addition, as we shall see, the excess neutrons could be used, again by means of new artefacts, in order to transmute the long-lived radioactive elements into stable species.

(4) Links to military applications: The critical mass of the plutonium from a LWR is only some 30% larger (6 kg) than that of bomb-grade ^{239}Pu. An ill-minded group of individuals — particularly if nuclear power becomes widespread in developing countries, whose rapid evolution makes them intrinsically more unstable than developed countries — may realize quite terrifying devices of the Nagasaki type.

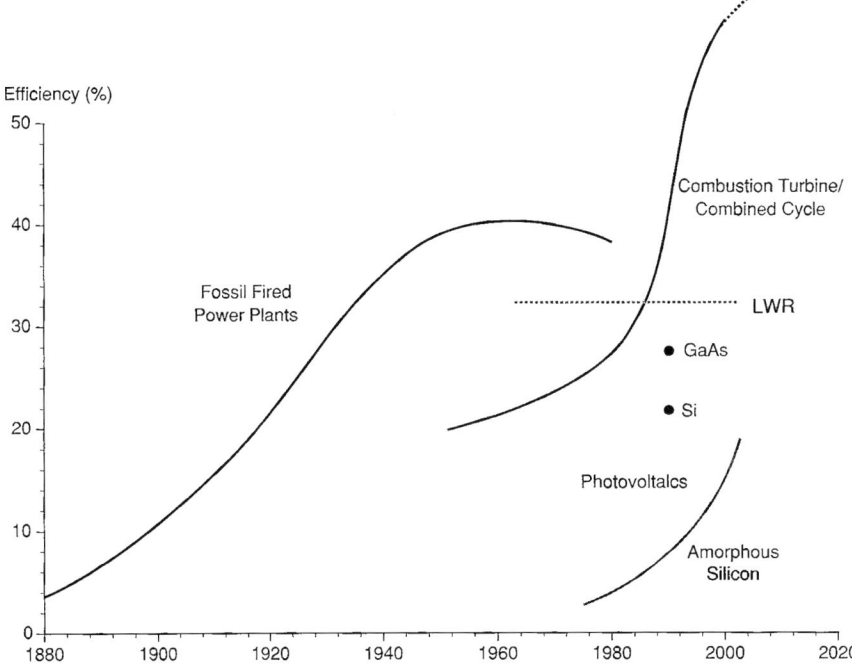

Figure 8. Thermodynamical efficiency of power stations.

New technologies must be made extremely robust in order to forestall deviated, military or terrorist proliferation.

(5) It is well known that thermodynamical efficiency, that is, the fraction of thermal energy that actually ends up in electricity, is temperature-dependent (Carnot), and it is about 33% for LWRs, related to their actual level of technological development in the late 1960s. Since then, enormous progress has been made in this field (Figure 8), and today gas-fired power stations can deliver efficiencies approaching double that of nuclear power stations. This means not only that the productivity and, consequently, the income for a given installation is doubled, but also that the caloric waste energy dissipated in the environment is halved. In order to keep their competitive edge, nuclear energy installations will have to increase their operating temperature substantially, and that means abandoning the saturated steam option of classic LWRs.

To conclude, in order to harness realistically the immense potential energy inside nuclei, very tough revival conditions must be satisfied, which, in turn, will inevitably demand new methods and new ideas. In short, we must wipe out three words from the list of popular concerns: (1) Hiroshima, (2) Chernobyl and (3) the geological repository (Juca Mountain, Hanford). In addition, we must use a naturally abundant fuel far more efficiently in order to secure its wiser use and its practically unlimited resources. As will be illustrated in the next section, both fusion and accelerator-driven fission have a fighting chance of achieving this goal.

4. Toward a Renovated Scenario

As is well known, energy is released whenever low Z nuclei fuse or high Z nuclei fragment (packing fraction). This principle leads to two substantially different breeds of new devices, which will briefly be described and critically compared: fusion and the accelerator-driven Energy Amplifier (fission). Both methods hold the remarkable promises of $\eta = 1$ — that is, full combustion of the initial, natural fuel — and of virtually unlimited natural resources:

(1) Fusion, in its simplest form, consists of the magnetically confined burning of tritium (^3H) through the reaction:

$$^3_1H + {}^2_1H \rightarrow {}^1_0n + {}^4_2He + 17.6 \; MeV$$

The unstable tritium ($t_{1/2} = 12.33$ y) is produced by 'breeding' from lithium, using the produced neutron:

$$^6_3Li + {}^1_0n \rightarrow {}^4_2He + {}^3_1H + 4.8 \; MeV$$

Additional ^3_1H, which is needed to compensate inevitable losses, comes from the (fast) reaction $^7_3Li + {}^1_0n \rightarrow {}^4_2He + {}^3_1H + {}^1_0n$, in which the neutron is not destroyed. In this way we can achieve a breeding equilibrium, namely, a situation in which the amounts of ^3_1H produced and burnt are the same. The main shortcoming of this reaction, which is presumably the easiest to achieve, is that the bulk of the produced energy is carried by the fast (14 MeV) neutron, which, through secondary interactions, produces a considerable amount of activation in the reactor's structure.

(2) More advanced fusion reactions promise less radioactive activation. A different reaction would be possible with an initial deuterium–helium 3 mixture:

$$^3_2He + {}^2_1H \rightarrow {}^4_2He + {}^1_1p + 18 \; MeV$$

in which, however, some neutrons (6%) are produced in deuterium–deuterium collisions $^2_1H + ^2_1H \rightarrow ^3_2He + ^1_0n + 3.27\ MeV$. The main shortcoming of this reaction is the poor availability of ^3_2He. The best way proposed so far to gather this fuel is to bring it from the moon, where it has accumulated as a result of the solar wind. It is hard to believe that thousands of tons of fuel could be brought back to earth in an economically convincing fashion.

(3) One of the ultimate advantages of fusion, as compared, for example, with fission, is that there are several exothermic reactions that produce no neutrons, neither directly nor indirectly through secondary reactions. Since neutrons are the primary sources of activation, this makes the reaction inherently 'clean.' It is probably in this way that an ultimate nuclear energy will eventually be exploited in a very far-fetched future, excluding for the moment the possibility of a 'cold fusion.' The simplest reaction of this kind is $^1_1p + ^{11}_5B \rightarrow 3[^4_2He] + 8.78\ MeV$, which unfortunately is known not to 'ignite' in a magnetically confined device[10] (Tokamak) and probably also does not ignite under conditions of inertially confined fusion, driven by lasers or particle beams. Note that this reaction does not produce any gammas or neutrons. Both hydrogen and $^{11}_5B$ (81% of natural boron) are extremely abundant and easily obtained. More exotic devices are under study to exploit this formidable asset.

(4) Coming to fission, the accelerator-driven Energy Amplifier (EA) (Figures 9(a), 9(b) and 9(c)) is based on the fission reaction (FF: fission fragments):

$$^{233}_{92}U + ^1_0n \rightarrow 2.53[^1_0n] + 2FF + 200\ MeV,$$

driven by very fast ($\leq 1\ MeV$) neutrons from a high-energy accelerator. Just as in the case (1) of fusion, the $^{233}_{92}U$, which does not exist in nature, is bred from natural thorium by the reaction induced by secondary neutrons:

$$^{232}_{90}Th + ^1_0n \rightarrow ^{233}_{91}Pa + \gamma \xrightarrow{\beta - decay\ (27\ days)} ^{233}_{92}U + ^0_{-1}e.$$

As in case (1), a breeding equilibrium is reached, in which the amounts of $^{233}_{92}U$ produced and burned are equal. The EA also can burn the additional elements that are produced by $^{233}_{92}U$ capturing neutrons (5%

---

[10]The energy irradiated by the plasma because of electron Bremsstralhung exceeds the energetic yield of the reaction; hence the plasma will not ignite.

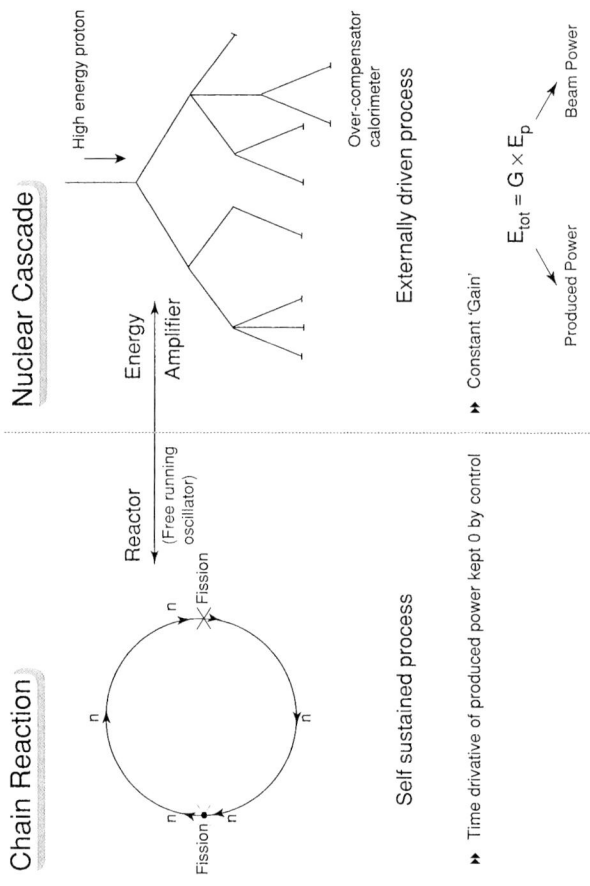

Figure 9(a). The general principle of accelerator-driven energy generation.

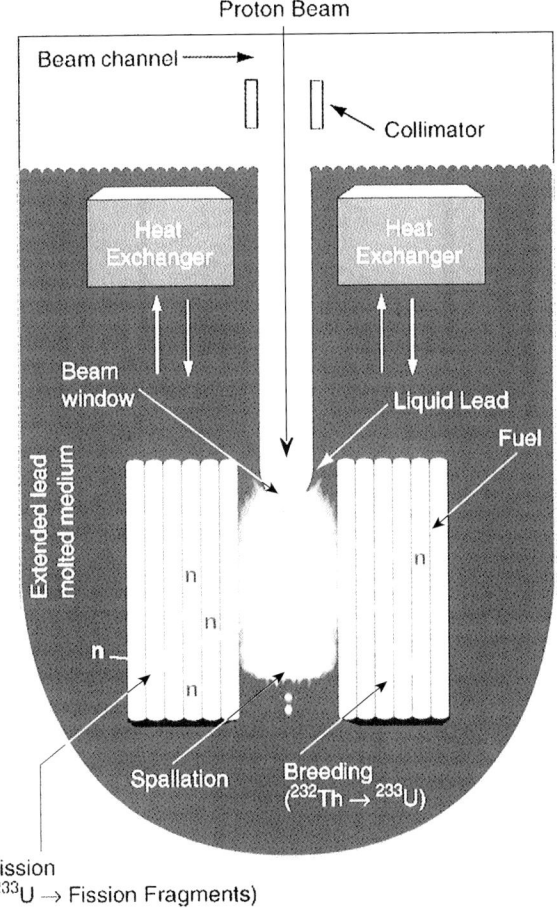

Figure 9(b). Diagram in principle of the Energy Amplifier.

of fissions) completely, with the process $^{233}_{92}U + ^{1}_{0}n \rightarrow ^{234}_{92}U + \gamma$ and the subsequent reactions, in a secular equilibrium with the main ones. Therefore, in contrast with LWRs, the EA achieves complete burn-up by fission of the initial $^{233}_{92}Th$, and therefore $\eta \approx 1$. Thus, the only 'waste' left is fission fragments, which have a strong but not very long-lasting activity.[11]

---

[11]The most offending long-lasting FFs can be easily transmuted into stable elements by means of the 'spare' neutrons from the main cycle.

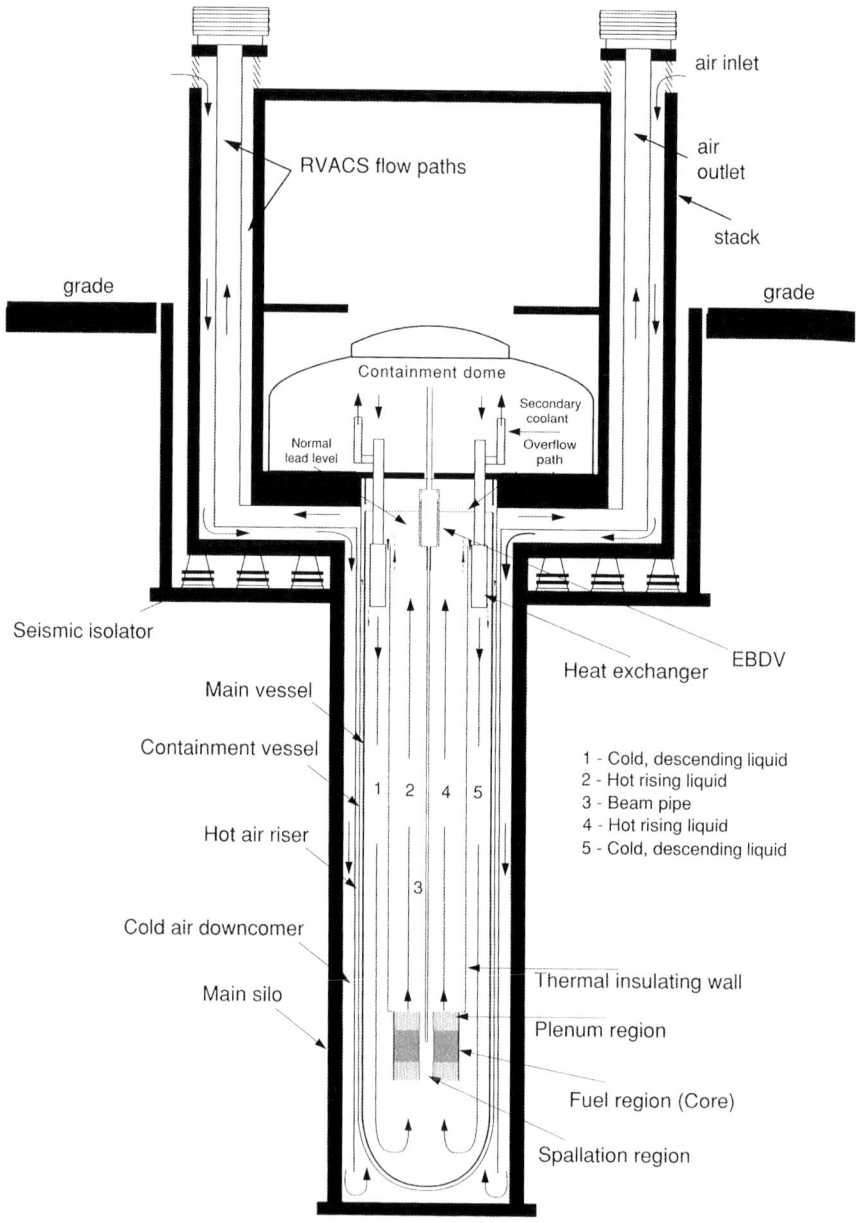

Figure 9(c). Practical layout of an EA.

(5) Though not intended for primary energy production, a new method has been derived from (4) in which the EA is used as an 'incinerator' of the most offending waste of the LWR, thus — at least in the transitional phase — solving a major problem of nuclear power as presently constituted. The method has the advantage of producing a substantial amount of energy in the process, thus paving the way for an economically rewarding activity. The device can also burn the military plutonium surplus, not a negligible point considering that the Cold War left us with several hundred tons of plutonium. The potential energy contained in 250 tons of military plutonium is the same as that of 3,500 millions of barrels of oil, roughly the yearly production of Saudi Arabia. We can either burn it quietly away, recovering all this energy, or take the risk that it may all fall on our heads in a global thermo-nuclear war! Elimination of the surplus plutonium from LWRs — which will eventually amount to about ten times the existing quantity of surplus military plutonium and, as already mentioned, is potentially quite dangerous — presents an equally rewarding mission, quite apart from the huge amount of recovered energy.

A combined cycle between the (existing) LWRs with thermal neutrons ($\langle E \rangle \approx 5.0 \times 10^{-2}$) and EAs with fast neutrons ($\langle E \rangle \approx 5.0 \times 10^5$ eV) could convincingly achieve the ultimate elimination of the major stockpile of long-lived TRUs (i.e., plutonium) and of the most pernicious FFs by the end of the life span of the program. In this way, no major radioactivity heritage would be left for future generations to worry about, after the exploitation of the nuclear cycle has been completed. In this logic, a useful term of reference for what would be left behind is the activity of the initial natural fuel ores. Actually, these methods hold the even better promise of reducing the level of residual radiotoxicity to a much lower level, similar, say, to that of coal ashes for the same produced energy. No sensible person would propose a geological repository for coal ashes!

In addition, both the fusion and the fission devices listed above are non-critical devices in which meltdown has been rendered impossible. In both devices, a fraction f of the produced (electric) energy is recirculated, either to heat up the plasma or to run the accelerator. This fraction is equal to 25–30% for devices of type (1) — i.e., D-T magnetically confined fusion — and to 5–10% for the EA, type (4). In the following discussion, we shall limit our considerations to devices (1) and (4). Device (5) is essentially identical to (4), apart from its different choice of fuel.

5. Comparing Magnetically Confined Fusion
   with Accelerator-Driven Fission

The main motivation for the research and development of new sources of
energy from nuclei is that of *reconciling the inherent advantages of such
powerful and virtually unlimited energy sources with an environmentally
acceptable and safe new technology.* This has been the main thrust behind
fusion, and it explains why so many people have been working so hard and
so long to achieve it. The far less ambitious development of the accelerator-
driven Energy Amplifier stems from the same objectives. It is therefore
reasonable that the potentialities of both methods be compared and critically assessed.

The development of an Energy Amplifier represents a relatively modest extrapolation of today's accelerator and nuclear energy technologies
and so can be predicted relatively well. The assessment of future fusion
power plants, by contrast, requires extrapolation from present-day physics
and technology to a time many decades ahead. In particular, it requires
judgments regarding feasibility in the areas of plasma confinement, stability, exhaust and special materials. Nevertheless, magnetic fusion (MF)
projected performances[12] represent a precise term of reference and a
gauge by which the validity of the Energy Amplifier (EA) concept should
be evaluated.

There is no doubt that environmental and safety features should govern
any new development in the field of energy from nuclei. We have reached the
conclusion that on all the relevant points in this area, magnetic fusion (MF)
as presently conceived and the EA design offer comparable environmental
and safety features.

The reference level of radiotoxicity should be lowered to that of coal
burning, an obvious case where the geological repository of ashes is not
required, though tiny amounts of U and Th decay chains are present. In
both cases (Figure 10), after a suitable 'cool-down' period, the radioactive
'waste' reaches radiotoxicities comparable to or smaller than that of ashes
from coal burning for the same produced energy, thus virtually eliminating
the need for a geological repository. The cool-down time is general shorter
for MF, but in both cases does not exceed some 500 years, after which period
materials can safely be stored at a shallow depth. The volume of 'waste' is
much larger in the case of MF, which also yields a large stockpile of tritium

---

[12]They should indeed be taken as goals rather than as substantiated judgments that
a commercial plant could actually be built.

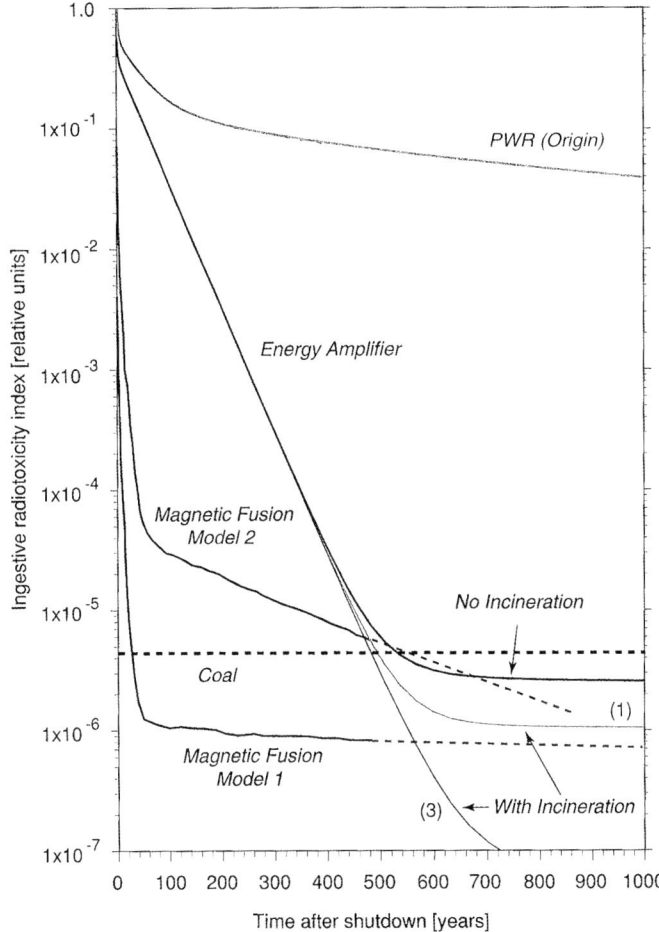

Figure 10. Comparison of residual radiotoxicity of EA, MF and LWRs, showing that of coal as a reference level.

that must be extracted and reprocessed on-line.[13] Both devices, after their very reasonable cool-down period, are enormously cleaner (5×10^{-5} for

---

[13]The bred tritium inventory is renewed every 100 hours, while in the EA the lifetime (1/turnover rate) of the ^{233}U is approximately 4 years. The toxicity of the total amount of tritium produced and handled is therefore more than two orders of magnitude greater than that of ^{233}U. In the case of the EA, the breeding process is an undisturbed step in the natural fuel evolution and takes place inside the solid metal. The produced fuel remains sealed inside the pins during the whole duration of the fuel cycle (5 years). Breeding in the EA is therefore a spontaneous process, well known in ordinary reactors, and it carries no new technological problem.

ingestive toxicity of waste after 500 years) than the present and future generations of LWRs.

Both the EA and MF exhibit an extreme robustness against any conceivable accident, always with benign consequences. In particular, the β-decay heat, comparable in both cases, is such that it can be dissipated passively in the environment. The much simpler structure, the excellent heat distribution over the volume ensured by the natural convection of the molten lead, the much smaller number of components and the smaller unit power of the EA (1.5 vs. 3.8 $GW_{thermal}$) are added advantages and in general ensure an even better safety margin.

A point of considerable importance is related to nuclear proliferation, namely, the possibility that some diversion of the main scope of the devices could permit the production of a sizeable ($\geq 10\,kg$) amount of bomb-grade fissile material. In both MF and the EA, the neutron flux is sufficiently high as to enable the production of ^{239}Pu, by simple irradiation of depleted uranium, or of pure ^{233}U with thorium.[14]

In the case of the EA, all controls are performed with the accelerator, and there are no active components (pumps, etc.) in the tank. The main EA tank can therefore remain inaccessible and kept sealed for the whole duration of the fuel cycle, typically five years. Direct access to the fuel can be allowed during maintenance periods only, in the presence of an international inspection team.

Therefore, both for the EA and for MF the most credible dissuasion will be by the very simple but effective method of sealing the tank and the shipping casks. For MF, however, access to and maintenance of the machine are much more frequent, and the need for an inspector's presence is much more demanding.

Under normal circumstances there should be no directly fissionable material[15] near the MF site. This is not so for the EA, though the fuel is sealed inside the tank. Periodically (after 5 years), the EA fuel will reach its burn-up limit due to radiation damage and the build-up of FF gases.

---

[14]In general, the fissile yield from thorium is smaller than that from uranium. However, the resultant ^{233}U, because it has a smaller spontaneous neutron yield than ^{239}Pu, can be used to construct simpler bombs, since a 'cannon' geometry can be used.

[15]However, tritium, although not yet considered in the Treaty on the Non-Proliferation of Nuclear Weapons (NPT), would be present in large amounts at the MF plant. It should be included in the provisions of the NPT as soon as fusion becomes a reality.

The fuel should then be reconditioned — not reprocessed — by the following means:

(1) Actinides are left untouched in the bulk of the fuel and topped up with additional thorium, to compensate for burning.
(2) In the process, the majority of the FFs are extracted by simple methods. Noble metals are mostly left in, while volatiles are naturally separated and rare earths extracted by oxidation.
(3) The fuel (metal) is recast and packaged with new cladding elements.

This fuel reconditioning procedure should cause no concern about proliferation risks, since contact fabrication, essential for the realization of nuclear explosives, is inhibited by the high radiation level of the mixtures of different chemical elements. Even if a chemical separation is attempted, the relatively large amounts of fissionable uranium are strongly (isotopically) contaminated by the very strong γ-emitter of the ^{232}U decay chain, which makes the extracted uranium inapplicable for military diversions.

The natural abundance of the materials used by MF and by the EA offer roughly the same energetic potentials. Referred to natural elements (lithium vs. thorium), the EA is more fuel efficient than MF by the factor of 4.12 in weight.[16] Lithium is estimated to be about seven times more abundant on the earth's surface than thorium (10^{-5} in relative abundance), but these differences are of little or no relevance, since both elements exist in sufficient quantities for millions of years of very intensive utilization.[17]

But the EA has no major technological barriers, is very likely to be much cheaper and to operate more smoothly than MF, and could be ready to take over nuclear energy production by the time the nuclear plants of the present generation reach the end of their lifetime.[18] Finally, EAs can definitively eliminate, by burning, the 'waste' from present nuclear plants and the surplus plutonium from warheads, vastly reducing the needs for geological storage of the existing and projected waste.

---

[16] Expressed in more practical units, 1 kg of natural lithium is equivalent to 3,500 barrels of oil, 1 kg of thorium to 15,000 barrels.
[17] Reasonable extrapolation of the thorium ores with a content larger than 2,000 p.p.m., the presently exploited uranium level, would indicate energy reserves of 4.5 million tWatt $\times$ year, corresponding to 2,200 centuries at twice the present world consumption, truly unlimited by the standards of human civilization.
[18] This is, in about 15–20 years from now, when today's plants will have been in operation for some 35–45 years.

From the standpoints of cost, simplicity of operation and reliability, the EA has a significant advantage over MF. It offers a realistic opportunity to produce emission-free energy at lower cost than fossils (Figure 11), for which it could therefore be a realistic substitution. In our view, the EA deserves consideration as a serious candidate for low-cost, clean nuclear energy. The pros and cons of the EA and of MF are listed in Table 2.

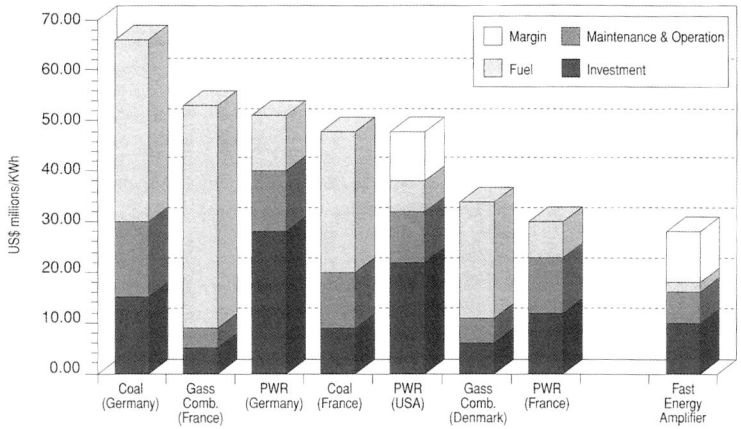

* From 'Projected Costs of Generating Electricity from Power Stations for Commissioning in the Period 1995-2000' (OECD Nuclear Energy Agency, 1989) and from IEPE-Grenoble.

Christian Roche, September 1996

Figure 11. Cost comparisons of different electricity-producing methods.

Table 2. Comparison of Magnetic Fusion with the Energy Amplifier.

Item	Magnetic Fusion (DEMO)	Energy Amplifier
Safety	not critical, no meltdown	not critical, no meltdown
Credibility	ignition unexplored	proven at zero power
Fuel	natural lithium	natural thorium
Fuel Availability	practically ∞	practically ∞
Chemistry of fuel	on line (tritium extraction)	every 5 y (regeneration)
Waste disposal	$\leq$ coal ashes after 500 y	$\leq$ coal ashes after 500 y
Operation	very complicated	very simple
Technology	major unknown	no major barrier
Prolif. resistance	good, but open geometry	excellent: sealed fuel tank
Cost of energy	presumed larger than fossil	presumed less than fossil
Realization time	long ($>$ 50 years)	short ($<$ 10 years)

6. Conclusions

Let me end where I began. Can the environment tolerate the emissions burden that ten billion energy-thirsty people will impose? That is too tough a question to answer conclusively, and I would add that it is not up to scientists to solve all the problems — the foremost of which is population — by themselves. However, *innovation is the most powerful of our renewable resources*: We are in the process of achieving a better understanding and better answers on the issues surrounding climate concerns, nuclear waste and the technological opportunities that will extend our finite natural resources.

It is evident that the world, though more crowded, will also become smaller, its peoples linked in real time by language as well as by bytes and pictures. The world's aspirations will grow. Governments will be under increasing pressure to provide new opportunities, to anticipate aspirations and to respond to new needs. Science and technology must be channeled toward the peaceful achievement of these goals.

Prof. Carlo Rubbia delivered the Albert Einstein Memorial Lecture in 1998.

Supramolecular Chemistry: From Molecular Information toward Self-Organization and Complex Matter

Jean-Marie Lehn

1. Introduction

> In the beginning was the Big Bang, and physics reigned. Then chemistry came along at milder temperatures; particles formed atoms; these united to give more and more complex molecules, which in turn associated into organized aggregates and membranes, defining primitive cells out of which life emerged. (Lehn, 1995)

From divided to condensed, organized, living, and up to thinking matter, the universe has evolved toward a progressive complexification of matter, through a process of self-organization (Eigen, 1971; Yates, 1987; Lehn, 2002a and 2002b) under pressure of information.

How matter becomes and has become complex is the most fundamental question posed to science, in that it addresses the issue of how (and why) the evolution of the universe has given rise to an organism capable of asking this very question and of generating the means to answer it, by creating science.

A parallel — if not more than that — may be drawn between structure formation on the grand scale of the universe and that which occurs on the level of molecular matter. The self-organization of the universe is a consequence of the operation, at very early times, of gravitational forces on initial inhomogeneities in density or in expansion rate (Rees, 2003; Coles, 2001; Ferreira, 2003). The self-organization of molecular matter, non-living and living (Eigen, 1971; Yates, 1987; Lehn, 2002a and 2002b) may be understood as resulting from electromagnetic forces generating and operating on an infinite diversity of possible structural combinations. Cosmic self-organization is thus due to gravitation, while molecular self-organization is due to electromagnetic interaction (Figure 1).

SELF-ORGANIZATION

of the UNIVERSE   of MOLECULAR MATTER

through   through
GRAVITATIONAL   ELECTROMAGNETIC
FORCES   FORCES

COSMIC STRUCTURE   ORGANIZED / LIVING /
THINKING MATTER

Figure 1. Cosmic self-organization and molecular self-organization.

Understanding self-organization, one may claim, is thus more basic than any other scientific endeavor, even one as fundamental, say, as general relativity or quantum theory, for it concerns the question of how the evolution of the universe has come to generate an entity capable of creating general relativity and quantum theory in the first place, an entity capable, by way of a radical shortcut, of interrogating the universe out of which it is born!

Chemistry, as the science of the structure and transformation of matter, has a major role to play in this context. It lies at the core of the biological world, the highest level of complex matter as we know it. Before biological evolution, spontaneous chemical evolution took place, performing selection on molecular structural diversity through electromagnetic interactions implementing molecular information. Chemistry has developed from mastering the combination and recombination of atoms into increasingly complex molecules to the harnessing of intermolecular forces for the generation of informed supramolecular systems and processes.

2. From Molecular to Supramolecular Chemistry

Over the last 150 years, molecular chemistry has developed a very powerful arsenal of procedures for making or breaking covalent bonds between atoms in a controlled and precise way. It has implemented those procedures for the purpose of constructing novel, ever more sophisticated molecules and materials that exhibit a range of original properties of broad interest for basic and applied science.

Supramolecular chemistry goes beyond molecular chemistry based on the covalent bond, with the aim of developing highly complex chemical systems from components interacting via non-covalent intermolecular forces (Figure 2). Over the last thirty years or so, it has grown into a major field of investigation and has fueled numerous developments at its interfaces with biology and physics, leading to the emergence and progressive establishment of supramolecular science and technology (Lehn, 1995; Atwood et al., 1996). This intensive activity has generated a spate of reviews, books and scientific conferences. Two recent special publications provide expert viewpoints on topics that have received especial attention in recent years (Special Feature, 2002; Viewpoints, 2002). Building on these and on general presentations of earlier work (Lehn, 1988, 1990 and 1995), the present essay aims at profiling some of the conceptual advances that have occurred in supramolecular chemistry and extrapolating some conjectures, mainly on the basis of work performed in the author's laboratories.

3. The Molecular Information Paradigm

Supramolecular chemistry has paved the way for the implementation of the concept of molecular information in chemistry, with the aim of gaining progressive control over the spatial (structural) and temporal (dynamic) features of matter and over its complexification through self-organization (Eigen, 1971; Yates, 1987; Lehn, 1988, 1990, 1995, 2002a and 2002b). By means of appropriate manipulation of intermolecular non-covalent interactions, it has explored the storage of information at the molecular level, in the structural features of the molecules, and the retrieval, transfer and processing of that information at the supramolecular level, via interactional algorithms operating through molecular recognition events based on well defined interaction patterns (hydrogen bonding arrays, sequences of donor and acceptor groups, ion coordination sites, etc.). This has involved the design and investigation of more or less strictly preorganized molecular receptors of numerous types, capable of binding specific substrates with high efficiency and selectivity — that is, through processes of high information content.

On the basis of such developments, chemistry may now be seen as an information science, the science of informed matter, involving the ever-clearer perception, deeper analysis and more deliberate application of the

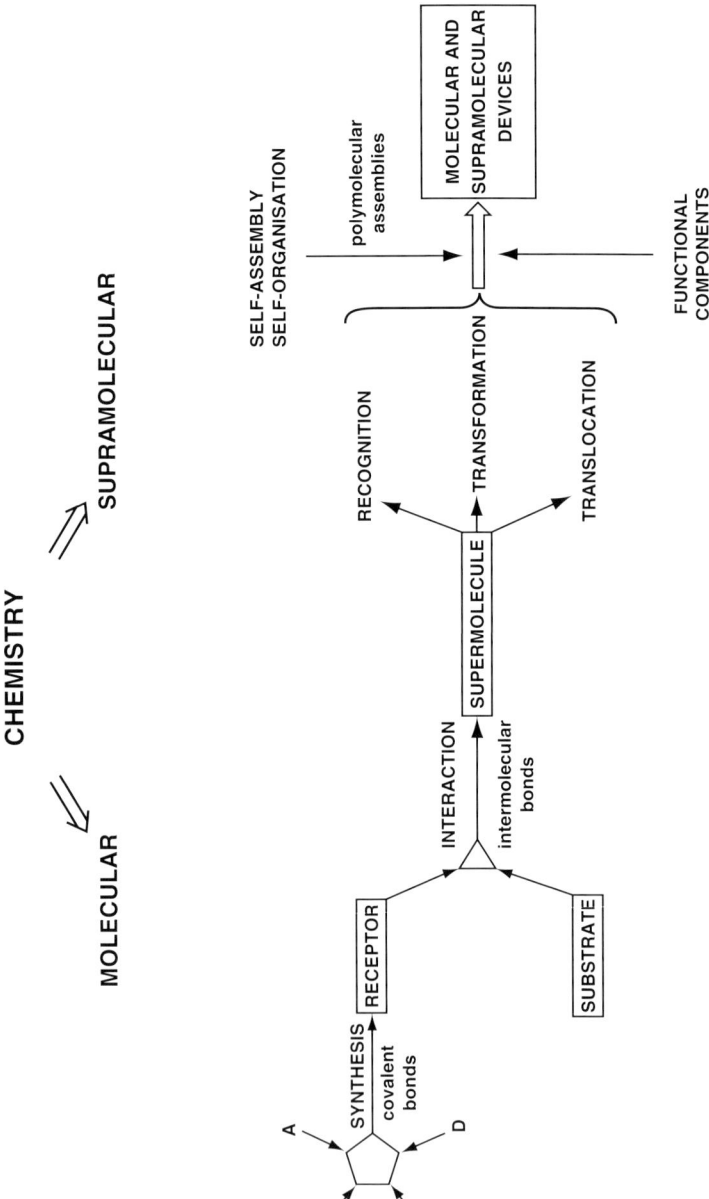

Figure 2. From molecular to supramolecular chemistry: molecules, supermolecules, molecular and supramolecular devices.

information paradigm in the elaboration and transformation of matter, thus tracing the path from merely condensed matter, to more and more highly organized matter, toward systems of increasing complexity. In chemistry as in other areas, the language of information is extending that of constitution, structure and transformation, as the field develops toward more and more complex architectures and behaviors. This process will profoundly influence our perception of chemistry, how we think about it and how we perform it.

Three main themes limn the development of supramolecular chemistry. The first, *molecular recognition*, relies on design and preorganization and implements information storage and processing. The second, the investigation of *self-organization* and self-processes in general, relies on design; it implements programming and programmed systems. The third, emerging phase introduces *adaptation* and *evolution*, based on self-organization by selection in addition to design, and implements chemical diversity and 'informed' dynamics.

4. Molecular Recognition, Catalysis and Transport

Supramolecular chemistry first harnessed preorganization for the sake of designing molecular receptors that effect molecular recognition, catalysis and transport processes (Lehn, 1988, 1990 and 1995; Atwood et al., 1996). It underwent rapid growth with the development of a wide variety of synthetic receptor molecules for the recognition — that is, the strong and selective binding, by means of various interactions (electrostatic, hydrogen bonding, Van der Waals, donor-acceptor) — of cationic, anionic or neutral complementary substrates of an organic, inorganic or biological nature. *Molecular recognition* implies the (molecular) storage and (supramolecular) retrieval of molecular structural information.

Many types of receptor molecules have already been explored, and still more may be imagined for the binding of complementary substrates of chemical or biological significance — for instance, for the development of substrate-specific sensors or for the recognition of structural features in biomolecules (nucleic acid probes, affinity cleavage reagents, enzyme inhibitors, etc.).

The combination of recognition features with reactive functions generates *supramolecular reagents and catalysts* that operate in processes involving two principal steps: substrate recognition followed by its transformation into products. Because of their relationship to enzymatic catalysis, they present protoenzymatic and biomimetic features. By nature, they are abiotic reagents that may perform the same overall processes as enzymes without following the same mechanistic pathways. More importantly, they may also effect highly efficient and selective reactions that enzymes do not perform. This represents a very important area for further development, with the aim of generating a range of reactive receptor molecules that combine substrate specificity with high reactional efficiency and selectivity. Much work remains to be done — work that should contribute very significantly to the understanding of chemical reactivity and its application in industrial processes.

Suitably modified receptors act as *carriers* for the selective *transport* of various types of substrates through artificial or biological membranes. Again, many further developments may be envisaged, such as the construction of selective membrane sensors or the transport of drugs through biological barriers, perhaps involving the design of artificial vectors for gene therapy and targeting if suitable target-selective recognition groups are introduced.

Recognition, reactivity and transport represent the three basic functional features of supramolecular species (Figure 2).

5. Functional Molecular and Supramolecular Devices

A further important line of development concerns the design of *functional supramolecular devices* based on photoactive, electroactive or ionoactive (and so forth) components, operating, respectively, by means of photons, electrons or ions. Thus, a variety of photonic devices based on photoinduced energy and electron transfer may be imagined (for references see Lehn, 1995; Atwood et al., 1996; Special Feature, 2002; and Viewpoints, 2002). Molecular wires, ion carriers and channels facilitate the flow of electrons and ions through membranes. Such functional entities represent entries into molecular photonics, electronics and ionics, which deal with the storage, processing and transfer of materials, signals and information at the molecular and supramolecular levels (for illustrations see Figure 3).

The use of external triggers (photonic, electronic, ionic, etc.) to induce molecular motions makes it possible to create functional devices that undergo mechanical processes. Such mechano-devices give access to a range of intriguing processes (such as shift registers and circular displacements) related in particular to the design of 'molecular machines' (Balzani et al., 2000; Stoddart, 2001; Collin et al., 2001).

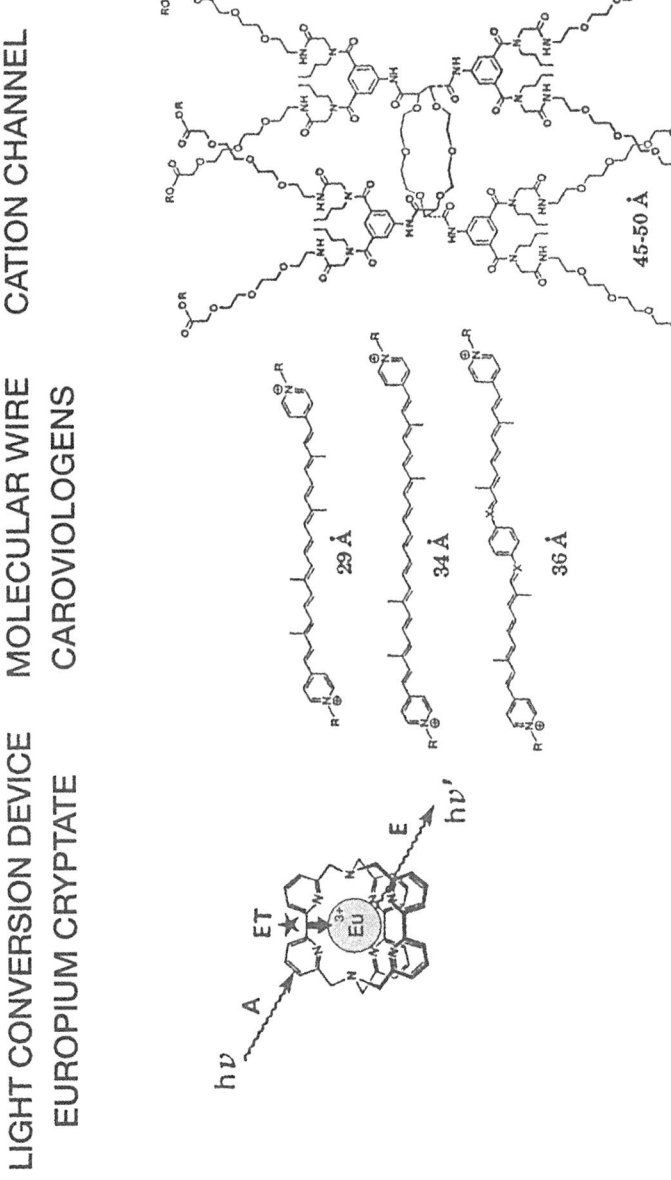

LIGHT CONVERSION DEVICE

EUROPIUM CRYPTATE

MOLECULAR WIRE

CAROVIOLOGENS

CATION CHANNEL

Figure 3. Diagrams (l to r) of photonic, electronic and ionic molecular devices (from work carried out in the author's laboratory).

For instance, very large amplitude contraction/extension motions can be generated by reversible metal ion binding, inducing interconversion between a helically wrapped and a linear stretched species (Barboiu and Lehn, 2002) (Figure 4).

A whole field at the interface of chemistry with physics, nanooptics, nanoelectronics and nanomechanics, wide open and barely explored, lies before us here. It presents such intriguing goals as the design of storage (battery), amplification, switching and rectification devices. *Semiochemistry*, the chemistry of molecular signal generation, processing, transfer, conversion and detection, touches upon both physico-chemical and biological signalization processes. It is implemented in such developments toward supramolecular technology as sensors and other optical or electronic devices, of interest for deriving logic functions and 'molecular computing' (Lehn, 1995, chap. 8; Atwood et al., 1996, Vol. 10; Balzani and Scandola, 1991; Jortner and Ratner, 1997; Gokel and Muhopadhyav, 2001; De Silva and McClenaghan, 2000; and Czarnik and Desvergne, 1997).

6. Self-organization by Design: Programmed Chemical Systems

The implementation of molecular information offers means of controlling the evolution of supramolecular species as they build up from their components. Thus, beyond preorganization lies the design of *programmed systems* in which self-organization results from the explicit manipulation of molecular recognition features so as to direct the build-up of these systems out of their components, and from the operation of supramolecular species and devices (Lehn, 1988, 1990, 1995, 2002a and 2002b; Atwood et al., 1996, Vol. 9; Whitesides et al., 1991; Lawrence et al., 1995; Leininger et al., 2000; Swiegers and Malefetse, 2000; Lindoy and Atkinson, 2000; Lindsey, 1991; Philp and Stoddart, 1996). Such programming involves molecular storage, in the components, of the information required for their assembly into a well-defined supramolecular entity, and supramolecular operation, based on specific recognition algorithms.

Understanding, inducing and directing self-processes is crucial to unraveling the progressive emergence of complex matter. Self-organization is the driving force that led up to the evolution of the biological world out of inanimate matter (Eigen, 1971; Yates, 1987). The inclusion of dissipative, non-equilibrium processes, as present in the living world, constitutes a major goal and challenge for the future.

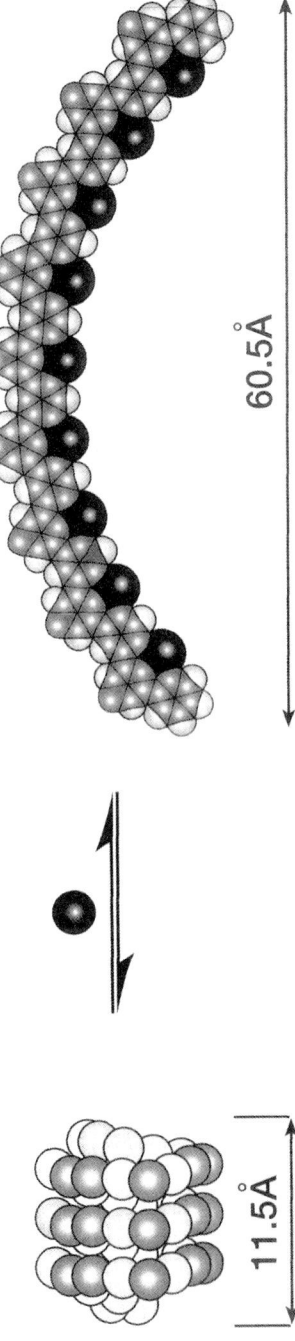

Figure 4. Dynamic devices: Large-amplitude reversible extension/contraction motion induced by interconversion by means of metal ion binding between a helically wrapped and an extended form of a molecular strand. Length change amplitudes by factors of up to 5 and 6 have been achieved.

More or less strict programming of the output species may be achieved, depending on the robustness of a given directing code (of the nature, for instance, of hydrogen bonding or metal coordination), that is, on the extent to which it is sensitive to internal factors (such as secondary metal coordination, van der Waals stacking, etc.) or external ones (such as concentrations and stoichiometries of the components, presence of foreign species, etc.). Sensitivity to perturbations, while limiting the range of operation, introduces diversity and adaptability (Lehn 1999a, 2002a and 2002b) into the self-organization process.

Self-assembly and self-organization have recently been achieved in numerous types of organic and inorganic systems (Lehn, 1988, 1990, 1995; Atwood et al., 1996, Vol. 9; Whitesides et al., 1991; Lawrence et al., 1995; Leinninger et al., 2000; Swiegers and Malefetse, 2000; Lindoy and Atkinson, 2000; Lindsey, 1991; Philp and Stoddart, 1996). By clever exploitation of metal coordination, hydrogen bonding or donor-acceptor interactions, researchers have achieved the spontaneous formation of a variety of novel and intriguing species, such as inorganic double and triple helices, termed helicates, cage compounds, catenanes, threaded entities (rotaxanes), and more.

The self-assembly of inorganic architectures of several types, effected by means of ligand design and the use of suitable coordination geometries that act as the assembling algorithm, may serve here as an illustration.

Double-stranded, triple-stranded and circular helicates, metal complexes of double and triple helical as well as circular helical architectures, are formed by the spontaneous assembly of two, three or five linear oligobipyridine ligands of a suitable structure into a double, a triple or a circular helix by means of the binding of specific metal ions displaying tetrahedral (Cu^I) or octahedral (Ni^{II} or Fe^{II}) coordination geometry. These species are represented by the trinuclear double and triple helicates (Lehn, 1988, 1990 and 1995; Atwood et al., 1996) and the pentanuclear circular helicate (Hasenknopf et al., 1997) shown in Figure 5.

Multiple component self-assembly leads to the spontaneous generation of multilevel cylindrical superstructures (Figure 6) from five to seven ligands of two different types and six to twelve Cu^I ions. This process represents the remarkable self-organization of a closed inorganic architecture from multiple components by means of the spontaneous and correct assembly, in one stroke, of eleven particles belonging to two types of ligands and one type of metal ion. Analogous cylindrical architectures of an even larger size, presenting three and four layered features, have been obtained using similar procedures (Baxter et al., 1999).

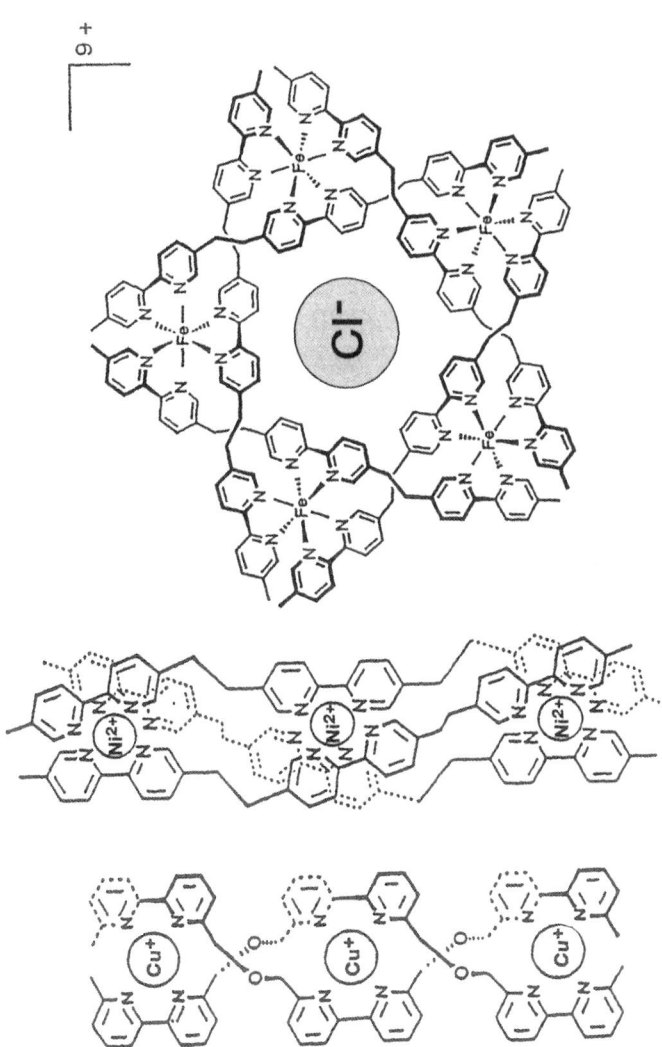

Figure 5. Double stranded, triple stranded and circular inorganic helices: a copper (I) double helicate (left); a nickel (II) triple helicate (middle); and a pentameric Fe (II) circular helicate containing a strongly bound chloride anion (right).

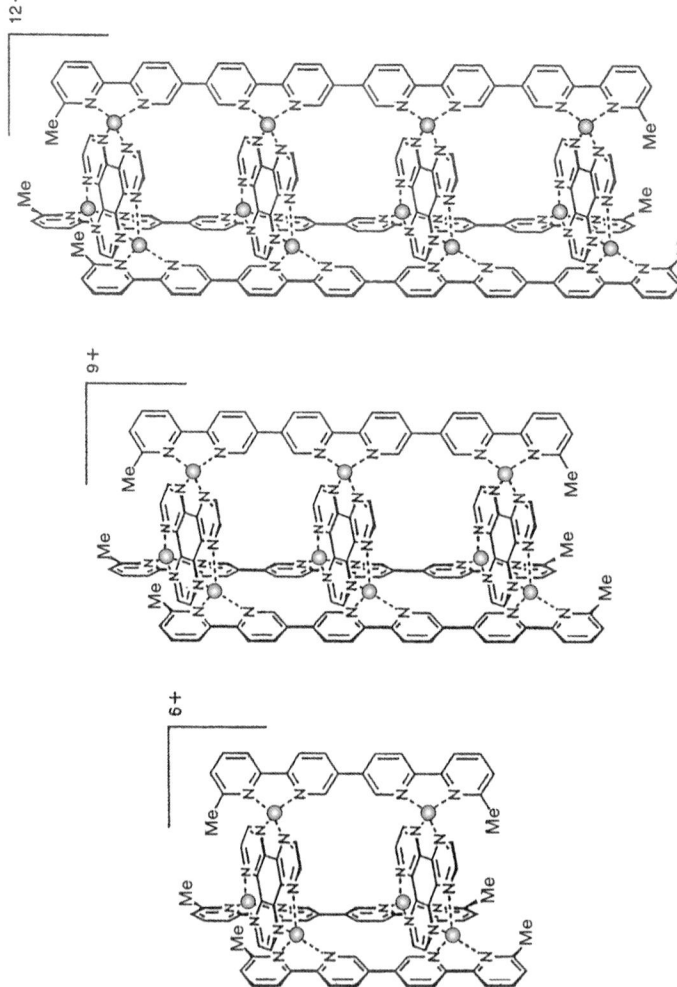

Figure 6. Multiple component self-assembly: multilevel cylindrical cage-like structures spontaneously generated from, respectively (left to right), five, six or seven ligands of two different types, and six, nine or twelve Cu^I ions.

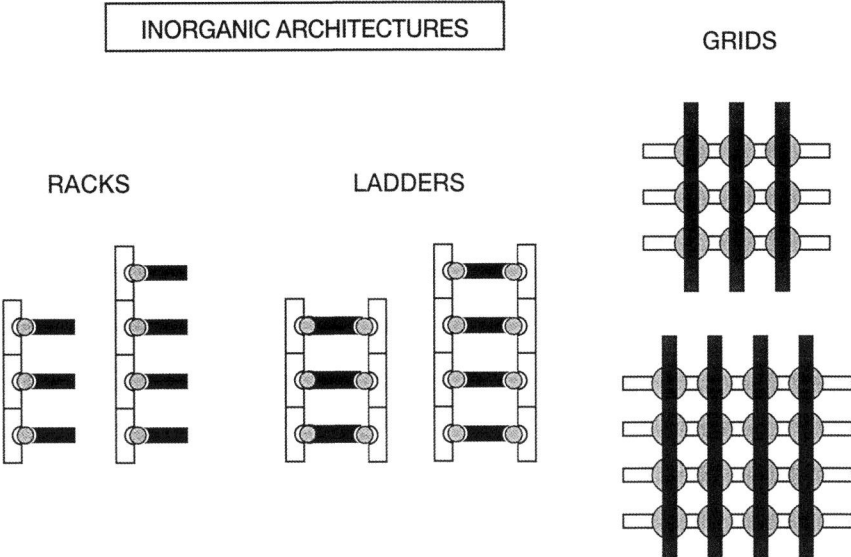

Figure 7. Inorganic architectures of rack, ladder and grid types generated by self-assembly.

Of particular interest are architectures of the rack, ladder and grid types (Figure 7). The latter are especially intriguing in view of their geometrical relationship to crossbar arrangements used in micro(nano)electronics. All three types of species have been generated by self-assembly.

Grid-like inorganic superstructures, such as the 3×3 and 4×4 grids shown in Figure 8, have been shown to self-assemble from their components: respectively, six ligands and nine Ag^{I} ions (Baxter et al., 1994), and eight ligands and sixteen Pb^{II} ions (Garcia et al., 1999).

Molecular-recognition directed self-organization, making use of hydrogen bonding, donor-acceptor and metal coordination interactions for controlling the processes and holding the components together, has given access to a range of supramolecular entities of truly impressive architectural complexity, such as would otherwise have been too difficult to construct (Lehn, 1995, 1988 and 1990; Atwood et al., 1996, Vol. 9; Whitesides et al., 1991; Lawrence et al., 1995; Leinninger et al., 2000; Swiegers and Malefetse, 2000; Lindoy and Atkinson, 2000; Lindsey, 1991; Philp and Stoddart, 1996), as well as to interlocked, mechanically linked compounds (Sauvage, 1999).

The operation of such instructed supramolecular systems fulfils the three levels of molecular programming and information input — *recognition,*

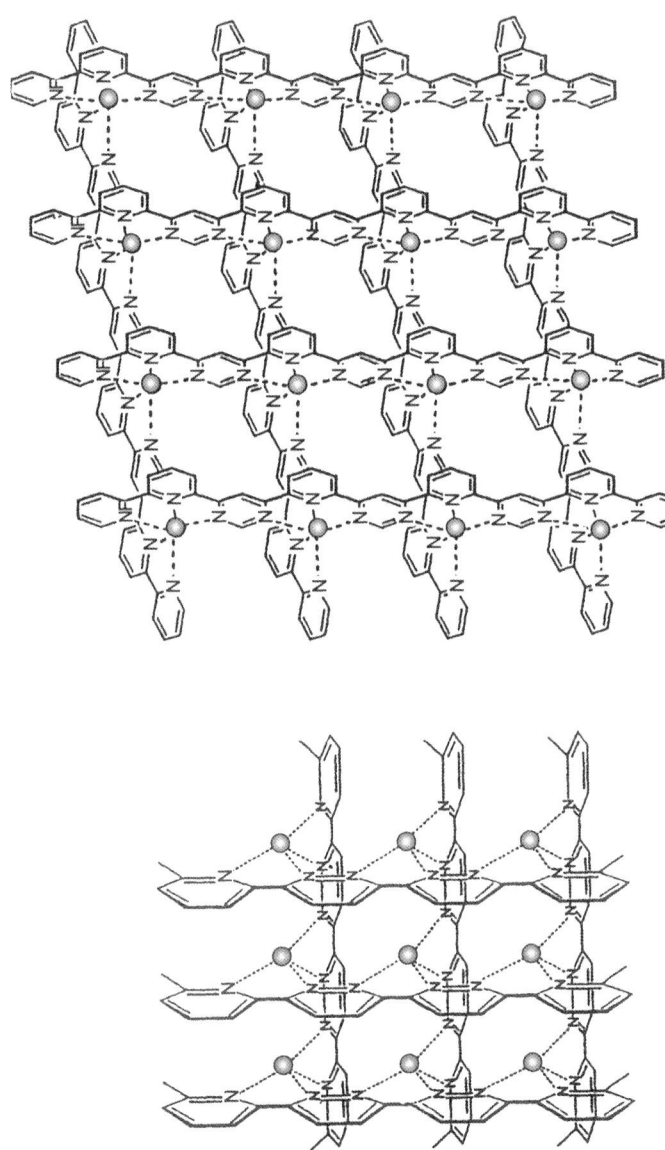

Figure 8. 3×3 (left) and 4×4 (right) grid-type inorganic architectures obtained by self-assembly from their components, respectively: six ligands and nine silver (I) ions; eight ligands and sixteen Pb (II) ions.

growth and *termination* — that determine the self-generation of a discrete supramolecular architecture. The steric and binding information contained in the ligand is read out by the metal ions following a given coordination algorithm. These processes represent progressive steps in the control of the self-organization of large, complex supramolecular architectures through molecular programming.

Chemically reactive self-organized entities are formed when the assembling brings together components bearing reactive functional groups. Through the appropriate disposition of specific subunits, they may be amenable to performing efficient and selective reactions and catalysis, and in particular may generate replication and self-replication processes (Robertson et al., 2000). The controlled self-organization of functional systems displaying reactivity and catalysis is crucial for the development of chemical systems of both structural and reactional complexity. It has played a key role in biological evolution (Eigen, 1971; Yates, 1987) and presents a major challenge to chemistry.

7. Self-Selection: The Instructed Mixture Paradigm

In a study of helicate self-assembly from a *mixture* of different ligands and different metal ions, it was found that only the 'correct' helical complexes (see Figure 5) were formed through *self-recognition* (Krämer et al., 1993) (Figure 9).

Self-selection with self-recognition occurs when the structural instructions are sufficiently strong. This process opens up a broader view toward the design of instructed components which, as mixtures, allow the controlled assembly of multiple well-defined supramolecular species, following specific information and interactional algorithms. The implementation of this 'instructed mixture' paradigm is crucial for the development of complex chemical systems, as witnessed by the build-up of organized species and the execution of highly integrated functions that take place side by side in the assembly and operation of the machinery of the living cell. One may venture to predict that this trend will represent a major line of development in chemistry in the years to come: the spontaneous but controlled build-up of structurally organized and functionally integrated supramolecular systems from a preexisting 'soup' of instructed components following well-defined programs and interactional algorithms.

Since it is a time-dependent process, self-organization also involves temporal information and may display kinetic control, as in the initial assembly

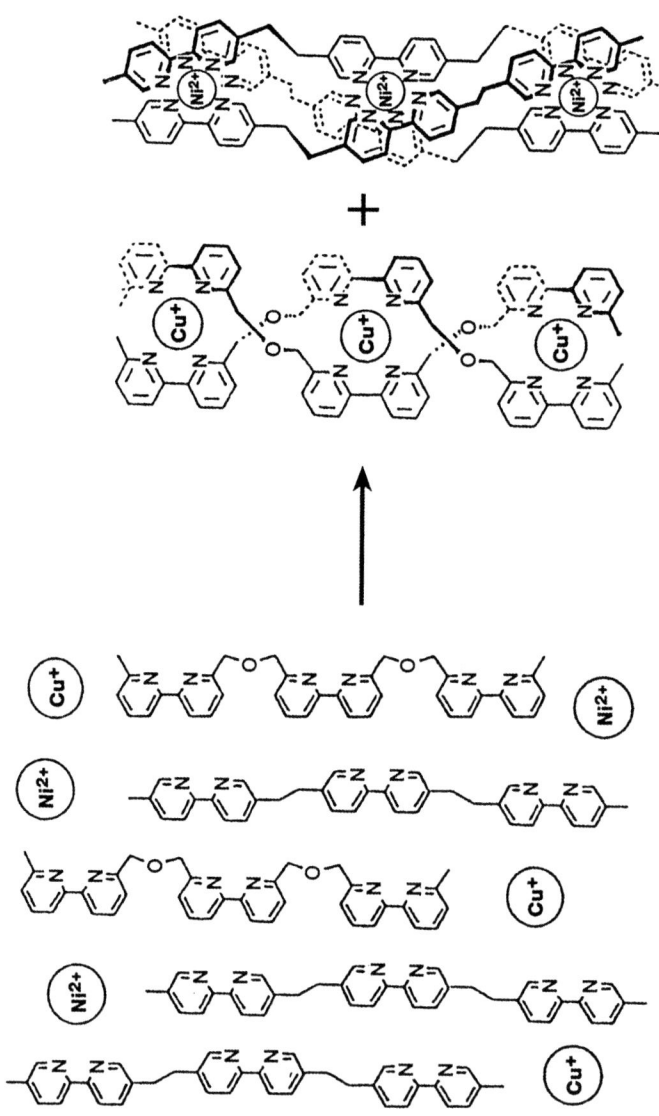

Figure 9. Parallel formation of a double helicate and a triple helicate, by self-selection with self-recognition from a mixture of two different suitably instructed ligands and two different types of metal ions presenting specific processing/coordination algorithms.

of a triple helical complex that evolves toward a circular helicate (Hasen-knopf et al., 1998).

Multilevel hierarchical self-organization, from primary, to secondary, to tertiary, etc., structures, enables the progressive generation of more and more complex systems in a sequential, temporally ordered fashion. Such is the case, for instance, in the formation of discotic liquid crystals by virtue of 'sector'-shaped components assembling into disks, which thereafter organize into columns (Suárez et al., 1998); and in the template-induced wrapping of molecular strands into helical disk-like objects, which then aggregate into large, supramolecular assemblies (Berl et al., 2000).

8. Multicode Programmed Systems

Beyond processes involving a single-code assembly program lie systems of higher complexity which operate in multimode fashion through the implementation of several codes within the same overall program, resulting in multiple self-organization processes (Lehn, 2000, 2002a and 2002b).

Thus, different metallo-architectures may be generated from the same ligand by using different sets of metal ions/coordination algorithms for reading the binding information, as was done to effect the generation of two different helicates from the same strand (Lehn, 1999a; Smith and Lehn, 1996), and the assembly of ligands containing two different subunits — coding, respectively, for the formation of a helicate and of a 2×2 grid-type complex (Funeriu et al., 2000) (Figure 10). Similarly, the differential processing of hydrogen-bonding information contained in a molecular strand may yield different supramolecular structures (Berl et al., 2000).

The multiple processing of the same ligand information by different interaction algorithms (for example, through the use of different sets of metal ions or of different H-bonding effectors) allows the controlled generation of different output architectures, resulting in the multiple expression of molecular information (Lehn, 2000). Such a one code/several outputs scheme, as opposed to the one code/one product mode, in principle also has significant implications for biology.

The combination of different recognition/instruction features in a molecular program opens a door to the design of self-organizing systems capable of performing molecular computation (Conrad, 1993; Rothemund, 2000). Recent studies have described the use of biomolecules and DNA-based protocols to solve computational problems (Adleman, 1994; Chen and Wood, 2000). An approach making use of specifically designed non-natural

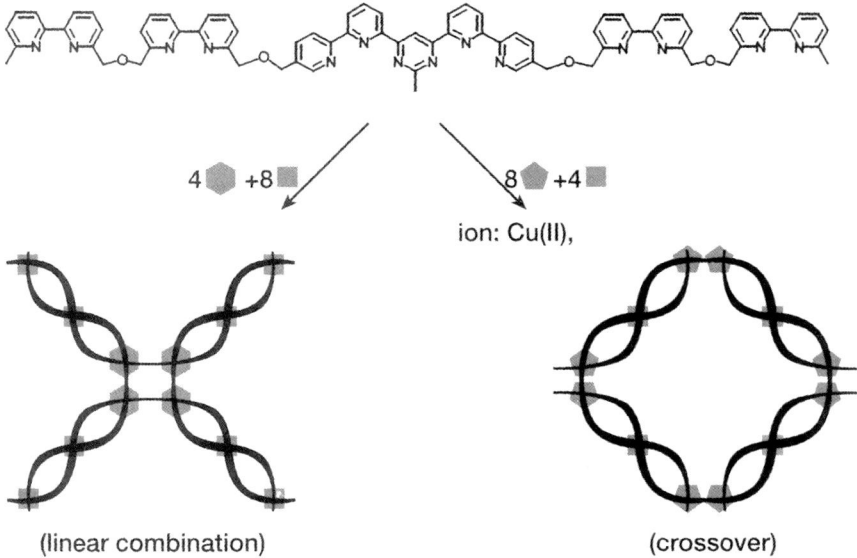

Figure 10. Generation of two different metallosupramolecular architectures by processing the molecular information of a ligand presenting two different recognition codes (for double helicate and 2×2 grid formation, respectively), by means of two different sets of metal ions of specific coordination geometries/algorithms.

components could provide higher diversity, better resistance to fatigue, and more compact, smaller size.

9. Self-Organization by Selection: Constitutional Dynamic Chemistry

Given the lability of the interactions connecting the molecular components in a supramolecular entity, supramolecular chemistry is by nature a dynamic chemistry. The reversibility of the associations allows for continuous change in the entity's constitution, by way of either internal rearrangement or exchange, incorporation and extrusion of components. Thus, supramolecular chemistry is a constitutional dynamic chemistry (CDC) generating constitutional diversity. It enables selection of a given constituent, made up of a well-defined set of components, out of the pool of compounds having all possible constitutions, under pressure of either internal factors (intrinsic stability of the species, as in helicate self-recognition; see Krämer et al., 1993) or external ones (interaction with species in the environment, as in anion binding by circular helicates; see Figure 5 and Hasenknopf et al., 1997).

CDC may also be operative on the molecular level, when the components of the molecular entity are linked by covalent bonds that may form and break reversibly.

Dynamic combinatorial chemistry (DCC), which has developed actively in recent years, is a specific expression of CDC (Lehn, 1999b; Cousins et al., 2000; Lehn and Eliseev, 2001). It rests on the dynamic generation of molecular and supramolecular diversity by means of the reversible connection of covalently or non-covalently linked building blocks, giving access to the full set of all combinations that may potentially exist. The addition of a receptor displaces the dynamic equilibrium toward the preferential formation of the best-binding constituent, in a target-driven selection of the fittest. This capability opens broad possibilities in various areas of science and technology, including the discovery of biologically active substances and of new materials.

CDC introduces a paradigm shift with respect to constitutionally static chemistry. The latter relies on design for the generation of a target entity, while CDC takes advantage of dynamic diversity to permit variation and selection. The implementation of selection in chemistry, both molecular and supramolecular, introduces a fundamental change in outlook. While self-organization by design strives to achieve full control over the output molecular or supramolecular entity by way of explicit programming, self-organization by selection utilizes the response of dynamic constitutional diversity to either internal or external factors in order to achieve adaptation in a darwinistic fashion.

Motional dynamic chemistry involves the design of supramolecular functional devices undergoing molecular motions triggered, as we saw above, by various physical or chemical stimuli (Balzani et al., 2000; Stoddart, 2001; Collin et al., 2001; Barboiu and Lehn, 2002).

10. Functional Supramolecular Materials

The properties of a material depend both on the nature of its constituents and on the interactions between them. Supramolecular chemistry may thus be expected to have a strong impact on materials science by virtue of its manipulation of the non-covalent forces that hold constituents together, leading up to the design of 'smart,' functional supramolecular materials and the control of their build-up and properties, generated by self-assembly from suitable units. The connections between the constituents being reversible, supramolecular materials may undergo assembly/disassembly/exchange processes. They are constitutionally dynamic materials (CDMs), in principle able to select their constituents in response to external stimuli or environmental factors, thus behaving as adaptive materials (Lehn, 1999a).

The combination of polymer chemistry with supramolecular chemistry defines a *supramolecular polymer chemistry* (Lehn, 1995; Ciferri, 2000; Brunsveld et al., 2001), in which molecular interactions (hydrogen bonding, donor-acceptor effects, etc.) and recognition processes are implemented to generate main-chain (or side-chain) supramolecular polymers by means of the self-assembly of complementary components (Figure 11). Supramolecular polymers are reversible CDMs, displaying constitutional diversity determined by the nature and variety of the different monomers. The exploration of supramolecular versions of the various species and procedures of molecular polymer chemistry gives access to a wealth of novel entities and functionalities (Ciferri, 2000; Brunsveld et al., 2001). Dynamic polymers, *dynamers*, may also be of a molecular nature if the connections between the monomers are based on reversible covalent bonds.

Molecular recognition may be used to induce and control self-organization in two and three dimensions so as to enable the supramolecular engineering of polymolecular assemblies, layers, films, membranes, micelles, gels and liquid crystals, both on surfaces or at interfaces and in the solid state (Desiraju, 1995). Vesicles are of special interest in this respect, since compartmentalization must have played a major role in the self-organization of complex matter and thus in the evolution of living cells and organisms. One may envisage the controlled build-up of architecturally organized and functionally integrated systems, leading up to the design of artificial cells and polyvesicular entities of tissue-like character for the performance of specific intra- and inter-vesicular processes (Menger and Gabrielson, 1995; Paleos et al., 2001). Recosomes — liposomes decorated with recognition groups — present such features as selective interaction with molecular films, aggregation and fusion (Marchi-Artzner *et al.*, 2001).

11. Self-Organization in Supramolecular Nanoscience and Nanotechnology

Nanoscience and nanotechnology are receiving great attention in view of both their intrinsic interest and their potential applications (Bard, 1994; Special Issue, 1999; R.F. Service, 2001). Again, supramolecular chemistry

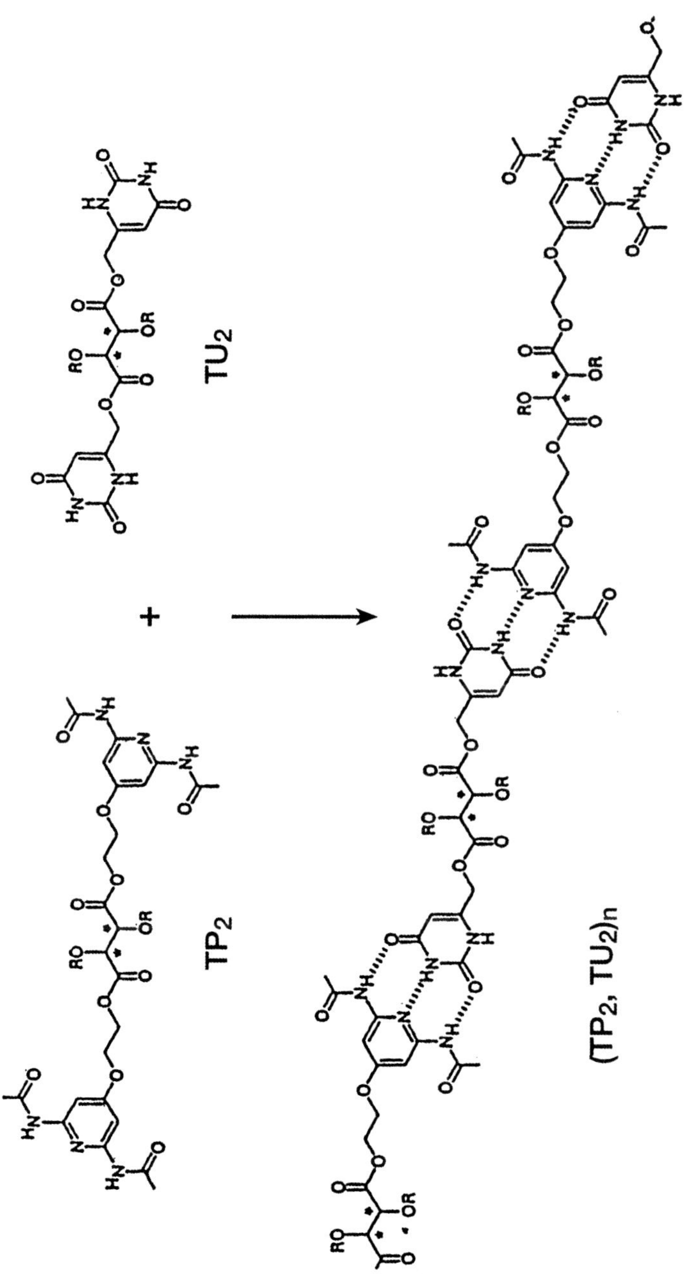

Figure 11. Supramolecular polymer chemistry: Mesophases and liquid crystalline polymers of a supramolecular nature have been generated from complementary components, amounting to macroscopic expression of molecular recognition.

and self-organization contribute a fundamentally novel outlook of potentially great impact. The generation of self-organized nanostructures (SONS) defines a supramolecular nanochemistry.

Self-organization offers molecular nanotechnology a powerful alternative to both top-down miniaturization and bottom-up nanofabrication. It strives for self-fabrication by means of the controlled assembly of ordered, fully integrated, connected operational systems generated by hierarchical growth, thus bypassing the need for tedious fabrication and manipulation procedures. It may utilize both design and selection, by virtue of its informed, dynamic and adaptive features, inspired by the integrated processes of biological systems. More and more powerful methodologies resorting to self-organization from instructed components should permit the generation of highly complex, functional supramolecular nanostructures. Of course, various combinations of self-organization and fabrication procedures may be envisaged and implemented at different stages.

Grid-type two-dimensional multimetallic inorganic architectures like those mentioned above (Figure 8) provide a prototypical illustration. Resembling the grids based on quantum dots that are of such interest in micro(nano)electronics, they may be viewed as consisting of *ion dots* of a size still smaller than that of quantum dots. Unlike the latter, they do not require nanofabrication but form spontaneously, by self-assembly. Such architectures may foreshadow multistate digital supramolecular chips for information storage in and retrieval from inscribed patterns that might be addressable by light, electrically or magnetically. Different states could, in principle, be characterized either by different local features at a given x, y coordinate, in ion dot fashion, or by specific overall optical or oxidation levels (see Lehn, 1995, chap. 10). Thus, 2×2 Co^{II}_4 (Ruben et al., 2003a) and Fe^{II}_4 (Breuning et al., 2000; Ruben et al., 2003b) grid complexes present, respectively, multiple electronic levels and triggered spin-transition processes that may be of interest for supramolecular electronics and spintronics. Addressing such entities may become possible via surface deposition (Semenov et al., 1999). It is noteworthy that molecular spin ladders display grid-type arrays (Rovira, 2000) and that cells presenting a four-dot 2×2 type architecture have been considered as potential components of quantum-dot cellular automata devices (Lieberman et al., 2002; Lent et al., 2003).

Reducing size to the nano-object and addressing it are admirable feats. In the long run, however, the goal is complex organization and collective operation rather than smaller size and individual addressing. It has become clear that the keyword of supramolecular chemistry is not size but information. Supramolecular species build up spontaneously from their components and accomplish complex tasks on the basis of encoded information and instructions. Thus, if, from the point of view of size, 'there's plenty of room at the bottom,' as the celebrated aphorism of Richard Feynman goes, then by means of supramolecular chemistry, 'there's even more room at the top!' (Lehn, 1995). The path is traced by self-organization, covering a full range of self-processes that determine the internal build-up, the functional integration, and the operation of the entity (self-selection, self-wiring), as well as its external connection to the environment (self-connection for addressing and sensing). Indeed, the most complex object around, the brain, builds up by self-organization and is self-wired and self-integrated, as well as self-connected through our senses!

12. Conclusion: Perspectives

The combined features of supramolecular systems — information and programmability, dynamics and reversibility, constitution and diversity — lead in the direction of adaptive/evolutive chemistry (Lehn, 1999, 2002a and 2002b). Adaptive chemistry implies selection and growth under conditions of time reversibility. It becomes evolutive chemistry when acquired features are conserved and passed on. Harnessing the power of selection for adaptation and evolution on the molecular scene is ushering a darwinistic era into chemistry. The ultimate goal is to merge design and selection in self-organization so as to accomplish self-design, with function-driven selection among suitably instructed dynamic species generating the optimal organized and functional entity, in a post-darwinian process.

Beyond programmed systems, the next step in complexity consists in the design of chemical 'learning' systems — systems that are not merely instructed but can be trained. The incorporation of time irreversibility implies passage from closed systems to open, coupled ones that are connected, spatially and temporally, to their surroundings.

Supramolecular chemistry provides ways and means for progressively unraveling the complexification of matter through self-organization, complexity being considered not just as the multiplicity of states but as the combination of multiplicity with interaction between and integration of states (Lehn, 1995). Together with the corresponding fields in physics and biology, it builds up a supramolecular science, leading to a science of complex matter — that is, of informed, self-organized, evolutive matter (Figure 12).

PERSPECTIVES ⇒ toward a SCIENCE of INFORMED MATTER

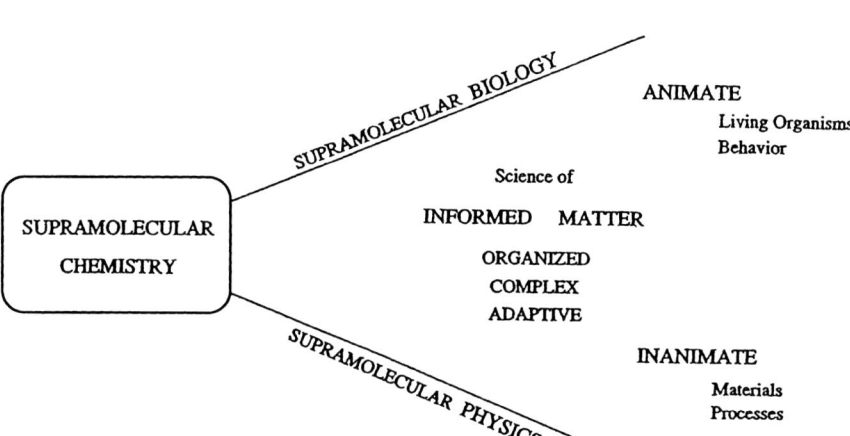

Figure 12. Supramolecular science as the science of informed matter: supramolecular chemistry at the interfaces of biology and physics.

The goal is to progressively discover, understand and implement the rules that govern the evolution of matter from inanimate to animate and beyond, in order ultimately to acquire the ability to create new forms of complex matter.

Prof. Jean-Marie Lehn delivered the Albert Einstein Memorial Lecture in 2003.

References

Adleman L.M. (1994) *Science* 266:1021–1024.

Atwood J.L., Davies J.E.D., MacNicol D.D., Vögtle F. and Lehn J.-M. (eds) (1996) *Comprehensive Supramolecular Chemistry*, Pergamon Oxford.

Balzani V., Credi A., Raymo F.M. and Stoddart J.-F. (2000) *Angew. Chem. Int. Ed.* 39:3348–3391.

Balzani V. and Scandola F. (1991) *Supramolecular Photochemistry*, Ellis Horwood, Chichester.

Barboiu M. and Lehn J.-M. (2002) *Proc. Natl. Acad. Sci.* 99:5201–5206.

Bard A.J. (1994) *Integrated Chemical Systems: A Chemical Approach to Nanotechnology*, Wiley, New York.

Baxter P., Lehn J.-M., Fischer J. and Youinou M.T. (1994) *Angew. Chem. Int. Ed. Engl.* 33:2284–2286.

Baxter P.N.W., Lehn J.-M., Kneisel B.O., Baum G. and Fenske D. (1999) *Chem. Eur. J.* 5:113–120.

Berl V., Krische M.-J., Huc I., Lehn J.-M. and Schmutz M. (2000) *Chem. Eur. J.* 6:1938–1946.

Breuning E., Ruben M., Lehn J.-M., Renz F., Garcia Y., Ksenofontoz V., Gütlich P., Wegelius E. and Rissanen K. (2000) *Angew. Chem. Int. Ed.* 39:2504–2507.

Brunsveld L., Folmer B.J.B., Meijer E.W. and Sijbesma R.P. (2001) *Chem. Rev.* 101:4071.

Chen J. and Wood D.H. (2000) *Proc. Natl. Acad. Sci. USA* 97:1328–1330.

Ciferri A. (ed) (2000) *Supramolecular Polymers*, Dekker, New York.

Coles P. (2001) *Cosmology: A Very Short Introduction*, Oxford Paperbacks.

Collin J.P., Dietrich-Buchecker C., Gavina P., Jimenez-Molero M.C. and Sauvage J.-P. (2001) *Acc. Chem. Res.* 34:477–487.

Conrad M. (1993) *Nanobiology* 2:5–30.

Cousins G.R.L., Poulsen S.A. and Sanders J.K.M. (2000) *Curr. Opin. Chem. Biol.* 4:270–279.

Czarnik A.W. and Desvergne J.-P. (eds) (1997) *Chemosensors for Ion and Molecule Recognition*, Kluwer, Dordrecht.

De Silva A.P. and McClenaghan N.D. (2000) *J. Am. Chem. Soc.* 122:3965–3966.

Desiraju G.R. (1995) *The Crystal as a Supramolecular Entity, Perspectives in Supramolecular Chemistry*, 2, Wiley, Chichester.

Eigen M. (1971) *Naturwissenschaften* 58:465–523.

Ferreira P.G. (2003) *Physics World* April:27–32.

Funeriu D.P., Lehn J.-M., Fromm K.M. and Fenske D. (2000) *Chem. Eur. J.* 6:2103–2111.

Garcia A.M., Romero-Salguero F.J., Bassani D.M., Lehn J.-M., Baum G. and Fenske D. (1999) *Chem. Eur. J.* 5:1803–1808.

Gokel G.W. and Mukhopadhyav A. (2001) *Chem. Soc. Rev.* 30:274–286.

Hasenknopf B., Lehn J.-M., Boumediene N., Dupont-Gervais A., Van Dorsselaer A., Kneisel B. and Fenske D. (1997) *J. Am. Chem. Soc.* 119:10956–10962.

Hasenknopf B., Lehn J.-M., Boumediene N., Leize E. and Van Dorsselaer A. (1998) *Angew. Chem. Int. Ed.* 37:3265–3268.

Jortner J. and Ratner M. (eds) (1997) *Molecular Electronics*, Blackwell, Oxford.

Krämer R., Lehn J.-M. and Marquis-Rigault A. (1993) *Proc. Natl. Acad. Sci. USA* 90:5394–5398.

Lawrence D.S., Jiang T. and Levett M. (1995) *Chem. Rev.* 95:2229–2260.

Lehn J.-M. (1988) *Angew. Chem. Int. Ed.* 27:89–112.

Lehn J.-M. (1990) *Angew. Chem. Int. Ed.* 29:1304–1319.

Lehn J.-M. (1995) *Supramolecular Chemistry, Concepts and Perspectives*, VCH, Weinheim.

Lehn J.-M. (1999a) in Ungaro R. and Dalcanale E. (eds), *Supramolecular Science: Where It Is and Where It Is Going*, Kluwer, Dordrecht, 287–304.

Lehn J.-M. (1999b) *Chem. Eur. J.* 5:2455–2463.

Lehn J.-M. (2000) *Chem. Eur. J.* 6:2097–2102.

Lehn J.-M. (2002a) *Proc. Natl. Acad. Sci. USA* 99:4763–4768.

Lehn J.-M. (2002b) *Science* 295:2400–2403.

Lehn J.-M. and Eliseev A. (2001) *Science* 291:2331–2332.

Leinninger S., Olenyuk B. and Stang P.J. (2000) *Chem. Rev.* 100:853–908.

Lent C.S., Isaksen B. and Lieberman M. (2003) *J. Am. Chem. Soc.* 125:1056–1063.

Lieberman M., Chellamma S., Varughese B., Wang Y., Lent C., Bernstein G.H., Snider G. and Peiris F.C. (2002) *Ann. N.Y. Acad. Sci.* 960:225–239.

Lindoy L.F. and Atkinson I.M. (2000) *Self-Assembly in Supramolecular Systems*, Royal Society of Chemistry, Cambridge, U.K.

Lindsey J.S. (1991) *New J. Chem.* 15:153–180.

Marchi-Artzner V., Gulik-Krzywicki T., Guedeau-Boudeville M.-A., Gosse C., Sanderson J.M., Dedieu J.-C. and Lehn J.-M. (2001) *ChemPhysChem* 2:367–376.

Menger F.M. and Gabrielson K.D. (1995) *Angew. Chem. Int. Ed.* 34:2091–2106.

Paleos C.M., Sideratou Z. and Tsiourvas D. (2001) *ChemBioChem* 2:305–310.

Philp D. and Stoddart J.F. (1996) *Angew. Chem. Int. Ed.* 35:1154–1196.

R.F. Service (2001) *Science* 293:782–785.

Rees M.J. (2003) *Our Cosmic Habitat*, Princeton University Press, Princeton.

Robertson A., Sinclair A.J. and Philp D. (2000) *Chem. Soc. Rev.* 29:141–152.

Rothemund P.W.K. (2000) *Proc. Natl. Acad. Sci. USA* 97:984–989.

Rovira C. (2000) *Chem. Eur. J.* 6:1723–1729.

Ruben M., Breuning E., Barboiu M., Gisselbrecht J.-P. and Lehn J.-M. (2003a) *Chem. Eur. J.* 9:291–299.

Ruben M., Breuning E., Lehn J.-M., Ksenofontov V., Renz F., Gütlich P. and Vaughan G.B.M. (2003b) *Chem. Eur. J.* 9:4422–4429.

Sauvage J.-P. and Dietrich-Buchecker C. (eds) (1999) *Molecular Catenanes, Rotaxanes and Knots*, Wiley-VCH, Weinheim.

Semenov A., Spatz J.P., Möller M., Lehn J.-M., Sell B., Schubert D., Weidl C.H. and Schubert U.S. (1999) *Angew. Chem. Int. Ed.* 38:2547–2550.

Smith V. and Lehn J.-M. (1996) *Chem. Commun.* 2733–2734.

Special Feature (2002) *Proc. Natl. Acad. Sci. USA* 99:4762–5188.

Special Issue on Nanostructures (1999) *Chem. Rev.* 99:1641–1990.

Stoddart J.-F. (ed) (2001) Special Issue on Molecular Machines, *Acc. Chem. Res.* 34/6:409–522.

Suárez M., Lehn J.-M., Zimmerman S.C., Skoulios A. and Heinrich B. (1998) *J. Am. Chem. Soc.* 120:9526–9532.

Swiegers G.F. and Malefetse T.J. (2000) *Chem. Rev.* 100:3483–3537.

Viewpoints (2002) *Science* 295:2400–2421.

Whitesides G.M., Mathias J.-P. and Seto C.T. (1991) *Science* 254:1312–1319.

Yates F.E. (ed) (1987) *Self-Organizing Systems*, Plenum, New York.

Chromatin and Transcription

Roger Kornberg

Genetic information in DNA is transcribed into RNA, which serves many purposes, including translation into proteins. Genomic DNA is packaged in nucleosomes, which impede the transcription process. The current article addresses the question of how this impediment is overcome. It is shown that nucleosomes are removed from promoter DNA, where transcription begins. Removal is accomplished by the disassembly of nucleosomes, through the action of chromatin remodeling complexes. A single nucleosome persists on the transcriptionally activated promoter at all times. Such nucleosome transactions are fundamental to all life processes, including cell differentiation and development, cell physiology and disease.

Glossary

Nucleosome	Fundamental unit of chromatin composed of an octamer of the basic histone proteins wrapped by a 146 base pair DNA stretch.
Promoter	Upper part of the gene that promotes/regulates the gene transcription
TATA box	The transcription initiation signal
ORF	Open Reading Frame, the part of the gene that codes the amino acids of the protein
RSC	Chromatin remodeling complex allows movement of nucleosomes along the DNA
Chaperone	A protein function in protein folding
Hybridization	Interaction of two complementary stretches of nucleic acids
Labeled probe (P)	Radioactive stretch of complementary DNA

Aharon Razin

My studies of eukaryotic gene transcription began with the discovery of the nucleosome, the basic unit of DNA coiling in eukaryote chromosomes. A decade later, Yahli Lorch and I found that the nucleosome represses transcription. How is this repression overcome? It has long been known that promoter DNA becomes more accessible to nuclease digestion following transcriptional activation. This led to the idea that nucleosomes are removed from promoters for activation. Then it emerged that the histone proteins of the nucleosome are still present in promoters following activation, but in an extensively modified form. Acetylation, methylation and other modifications, primarily of the so-called histone tails, are required for

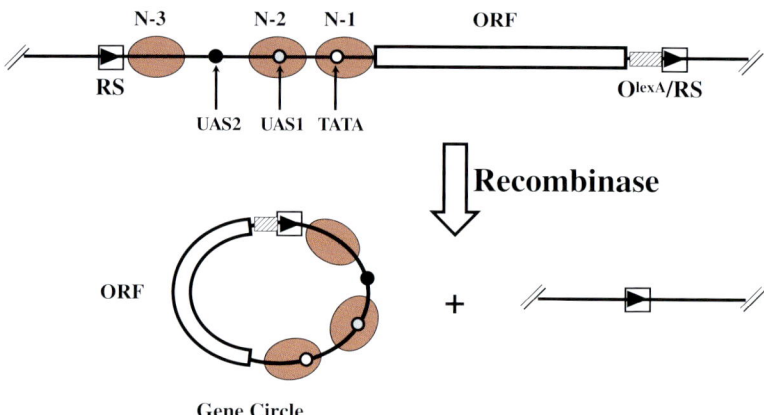

Figure 1. Site specific recombination at the *PHO5* locus: gene circle.

transcription. This led to the view that nucleosomes are not removed but rather reconfigured for transcription.

We wished to isolate the putative, reconfigured nucleosome, and to that end we took an approach introduced by Gartenberg and colleagues. We inserted recognition sites for the R recombinase on either side of the *PHO5* gene of yeast (Figure 1). Induction of the recombinase results in excision of the gene as a small circle, bearing 12 nucleosomes, three of which, as Wolfram Horz showed long ago, are located in the promoter region and are altered in some way upon transcriptional activation.

Any alteration in structure of the promoter nucleosomes upon transcriptional activation could be detected in a sensitive manner by measurement of the DNA topology. We found a linking difference of 1.85 between circles formed from repressed and activated *PHO5* genes (Figure 2). This difference might be attributable either to an altered state of the promoter nucleosomes, in which all three are partially unfolded, or to the complete unfolding of 1.85 of the original three nucleosomes.

We could distinguish between these possibilities by a quantitative analysis of the nuclease digestion of *PHO5* chromatin. Digestion with an enzyme such as micrococcal nuclease rapidly removes the DNA between nucleosome core particles and then more slowly degrades the core particles themselves (Figure 3). The process obeys a simple second-order rate equation, and it can easily be shown that the ratio of DNA remaining from activated and repressed promoters, as revealed by hybridization with a labeled probe, P,

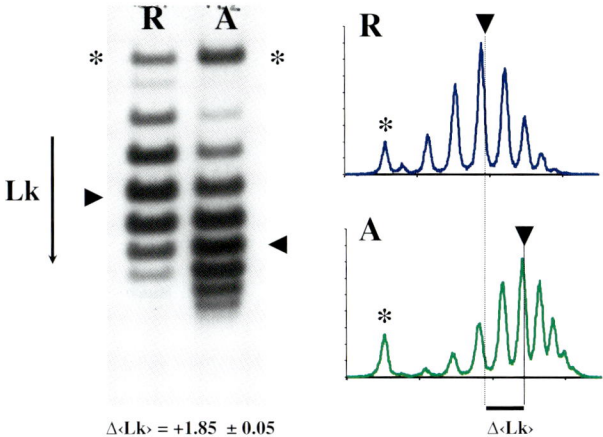

$$\Delta\langle Lk\rangle = +1.85 \pm 0.05 \qquad \Delta\langle Lk\rangle$$

Figure 2. Linking change associated with chromatin remodeling.

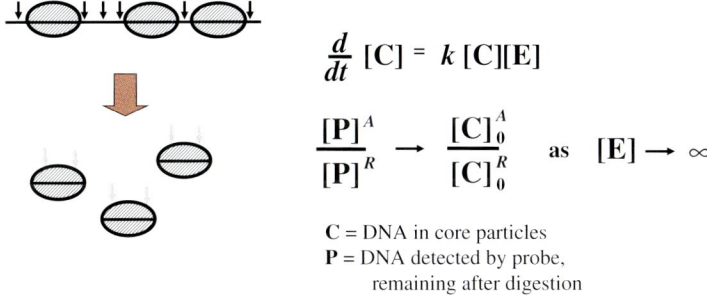

$$\frac{d}{dt}[C] = k\,[C][E]$$

$$\frac{[P]^A}{[P]^R} \longrightarrow \frac{[C]_0^A}{[C]_0^R} \quad \text{as} \quad [E] \longrightarrow \infty$$

C = DNA in core particles
P = DNA detected by probe,
      remaining after digestion

The ratio of DNA remaining from activated and repressed
promoters approaches the original ratio of core particles
at the promoters as digestion proceeds

Figure 3. Measurement of relative nucleosome number by limit nuclease digestion.

approaches the initial ratio of core particles, C_0, on the promoters, in the limit of complete digestion.

The results of experiments exhibit such asymptotic behavior, with a limiting value of 37% of the three core particles detected by the promoter probe retained upon activation, corresponding to 1.1 core particles retained (Figure 4).

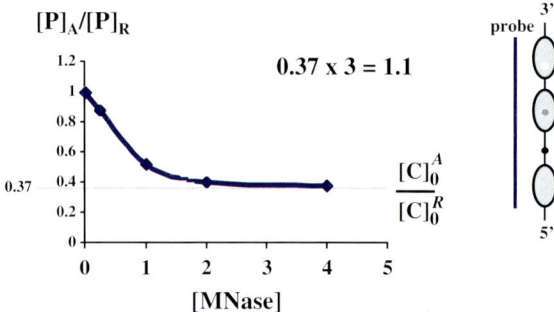

Figure 4. Number of nucleosomes retained on activated promoter.

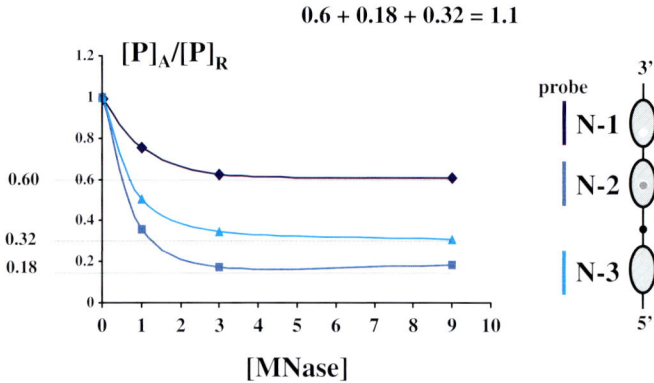

Figure 5. Nucleosomes are retained with different probabilities.

We repeated the experiment with individual probes for each of the three promoter nucleosomes (Figure 5). The behavior was again asymptotic, with the limiting values for retention of the individual core particles adding up to exactly the same value of 1.1 as from the previous analysis with a single promoter probe covering all three particles. The occupancies of the three promoter sites are 60%, 18% and 32% for N-1, N-2 and N-3. It is noteworthy that the TATA box and transcription start site are occupied by a nucleosome 60% of the time in the fully activated state of the *PHO5* promoter.

A third quantitative method corroborates this result. Here the activated promoter was cleaved with restriction endonuclease at sites flanking N-1,

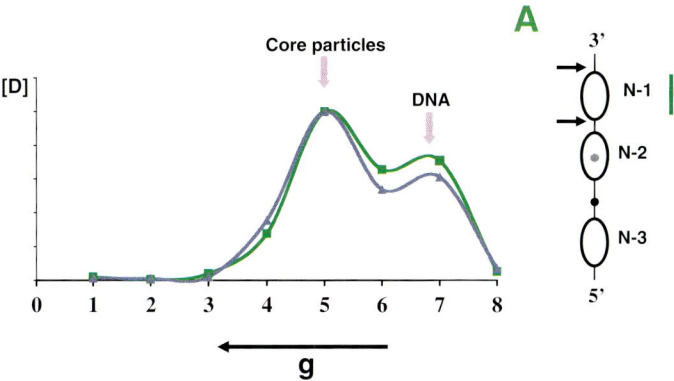

Figure 6. Sedimentation of N-1 released from activated promoter.

and the product of digestion was subjected to sedimentation analysis. The resulting sedimentation profile is in green; for comparison, the sedimentation profile of a 60–40 mixture of nucleosome core particles and naked DNA is shown in gray (Figure 6). By this method of analysis, the nucleosomes occupying position N-1 in the activated state are indistinguishable in structure from the canonical, well studied core particle.

Retention of a total of 1.1 nucleosomes, shown by limit nuclease digestion, corresponds to a loss of 1.9 of the original 3 nucleosomes, in close agreement with the results from topological analysis (Figure 7). The quantitative agreement from very different methods, along with additional evidence, leads to the conclusion that nucleosomes are removed from the transcriptionally activated promoter. So neither of the views prevailing in the past is entirely correct or entirely mistaken. Nucleosomes both are

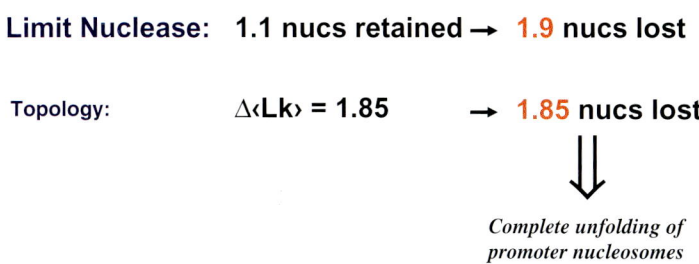

Figure 7. Complementary results of limit nuclease and topology analysis.

removed and are retained in unaltered form. The persistence of a nucleosome 60% of the time on the transcription start site implies a dynamic state
of the activated promoter, in which nucleosomes are continually removed
and reformed. Modified histones are presumably intermediates in the processes of removal and reformation.

If nucleosomes are removed, then what is the mechanism? There are
a number of reports of nucleosomes sliding away from the TATA box and
start site of transcriptionally active promoters. The alternative is that the
histone octamer is fully dissociated from the DNA (Figure 8).

We could again distinguish between these two possibilities by recourse
to chromatin circles (Figure 9). In our previous experiments, we activated
the *PHO5* gene and then induced the recombinase to form gene circles.
However, if we first form circles and then activate transcription, we can
answer our question about the mechanism.

The reason is that a circle is a closed domain, so if nucleosomes are
removed by sliding, their number on the circle will remain the same
(Figure 10). If, on the other hand, nucleosomes are removed by disassembly,
their number on the circle will be diminished.

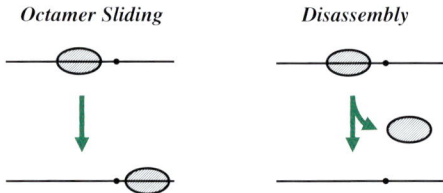

Figure 8. Mechanism of nucleosome removal.

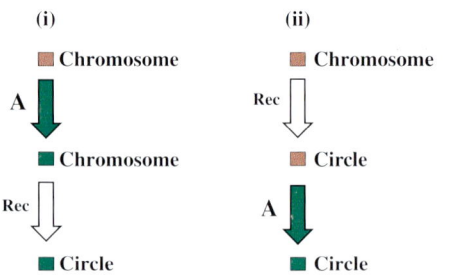

Figure 9. Reversing the order of activation and recombination.

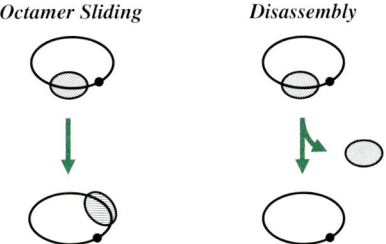

Figure 10. Mechanism of nucleosome removal.

1. Change in DNA supercoiling

$$\Delta <Lk>^{\text{activated -repressed}} = 1.50$$

2. Change in amount of nucleosomal DNA

1.44 nucleosomes lost
upon activation

⇓

*Histone octamers are
removed by disassembly*

Figure 11. Fate of *PHO5* nucleosomes upon transcriptional activation of promoter circles.

The result was that the number of nucleosomes was diminished (Figure 11). The amount of loss was slightly less than in the previous experiments, because the extent of activation was less, but results from topological analysis and limit digestion were again the same. We conclude that nucleosomes are removed by disassembly, not by sliding.

What is the cellular machinery for disassembly? We recently reported that the RSC chromatin remodeling complex could transfer histones to an abundant, ubiquitous histone chaperone protein, NAP1. This was revealed by the conversion of a labeled DNA from the mobility characteristic of a nucleosome to that of naked DNA in gel electrophoresis (Figure 12).

I'd like to return now to our original finding that nucleosomes are removed from the *PHO5* promoter in the course of transcriptional activation. This result raises two questions: What is the meaning of fractional values for the numbers of nucleosomes retained and lost? And why does any nucleosome remain, especially at the transcription start site? You might

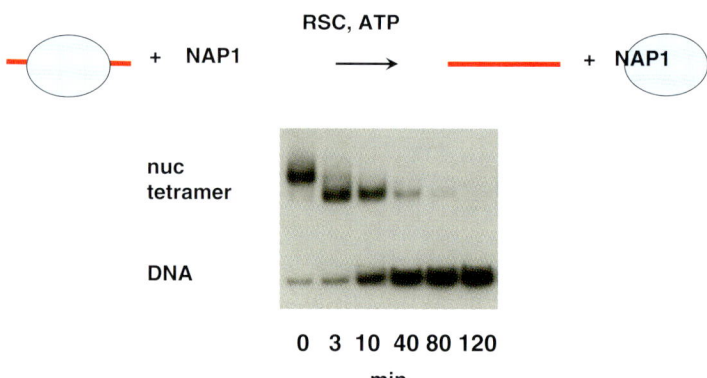

Figure 12. Histone octamer transfer to a chaperone protein by the RSC chromatin-remodeling complex.

imagine that fractional values represent averages over the many *PHO5* promoters in the chromatin circle preparation, with some circles retaining more and others fewer nucleosomes than the average. The question is statistical in nature, and we have addressed it by statistical methods. We performed a kinetic analysis, taking advantage of the slow time course of *PHO5* gene activation, which occurs over a period of about 12 hours. We excised promoter circles (Figure 13) at various times during the course of activation, and fractionated the circles on the basis of the number of nucleosomes

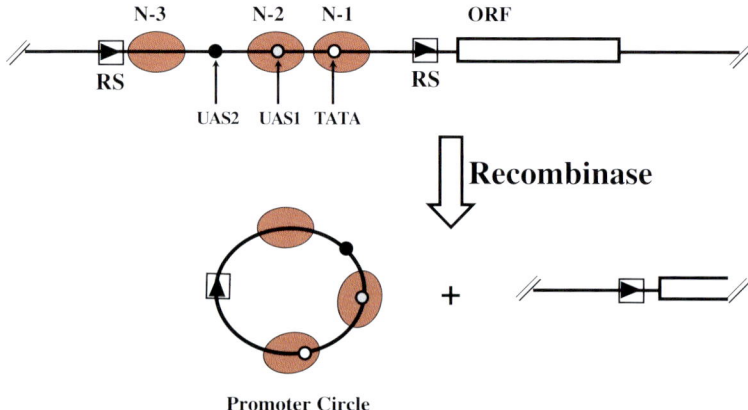

Figure 13. Site specific recombination at the *PHO5* locus: promoter circle.

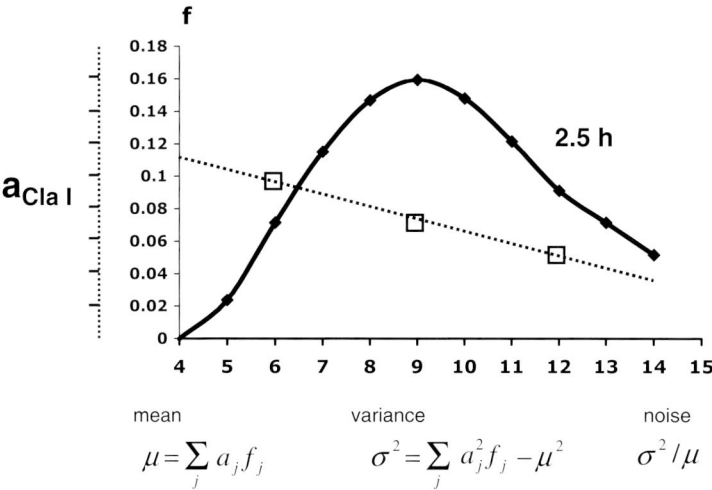

$$\mu = \sum_j a_j f_j \qquad \sigma^2 = \sum_j a_j^2 f_j - \mu^2 \qquad \sigma^2/\mu$$

Figure 14. Linear extrapolation of accessibility data.

by gel filtration — for example, after 2.5 hours of activation (Figure 14). We assessed the nucleosomal state of the circles in the gel filtration fractions by accessibility to cleavage with restriction enzymes — for example, Cla I, which cleaves a site in the middle of N-2. At early times, the circle population was homogeneous, with all circles in the repressed state, bearing three nucleosomes and inaccessible to Cla I digestion. At intermediate times, the circles were heterogeneous, with early-eluting circles bearing fewer nucleosomes and so more accessible to digestion than those eluting later from the column. At late times, the population was nearly homogeneous again, with most circles accessible to digestion. From the data, we could derive statistical quantities — for example, the mean of the distribution, given by the accessibility at a point in the elution profile, times the fraction of all circles at that point in the profile, summed over the entire profile. Similarly, we could derive the variance of the distribution and the statistical noise.

The time evolution of the variance of the circle distribution was thus derived (Figure 15). To interpret the behavior of the variance, we calculated what would be expected for possible pathways of chromatin remodeling. There are eight possible states of the promoter with 1, 2 or 3 nucleosomes, connected by a set of eight rate equations (Figure 16). Solution of the equations taking into account only nucleosome disassembly leads to the accumulation of all promoters in the naked promoter state, devoid of any

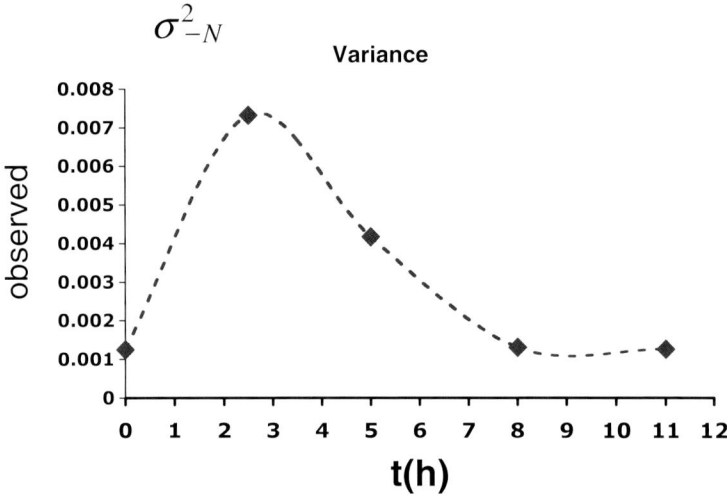

Figure 15. Statistics of *PHO5* promoter remodeling.

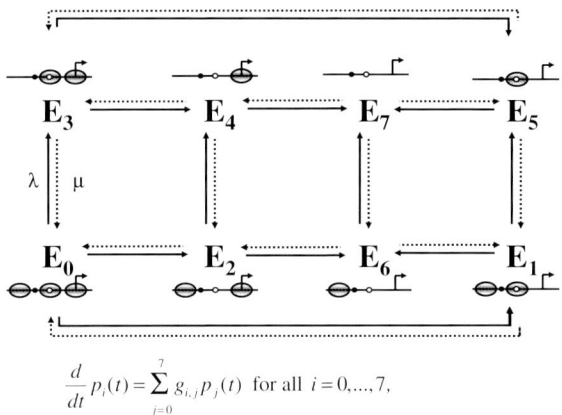

$$\frac{d}{dt}p_i(t) = \sum_{j=0}^{7} g_{i,j}p_j(t) \text{ for all } i = 0,...,7,$$

$p_i(t)$ is the probability of finding the promoter in state $E_{i \in \{0,...,7\}}$ at time t

Figure 16. Possible states of the promoter, with 1, 2 or 3 nucleosomes.

nucleosome, in conflict with the experimental evidence. This leads to the important conclusion that not only nucleosome disassembly but also nucleosome reassembly must occur in the transcriptionally activated state. You will recall that this same point could be inferred from the high occupancy of position N-1 and the transcription start site in the transcriptionally activated state.

If we now take into account reassembly and solve the rate equations with constants λ for disassembly and μ for reassembly, we arrive at a time evolution of the variance again inconsistent with the experimental results (Figure 17). The variance remains high, which is to say that the circle distribution remains very heterogeneous, even in the fully activated state. The reason is that to compensate for the formation of naked circles, devoid of all nucleosomes, and achieve an average of 1.1 nucleosomes, there must also be many circles bearing 2 nucleosomes.

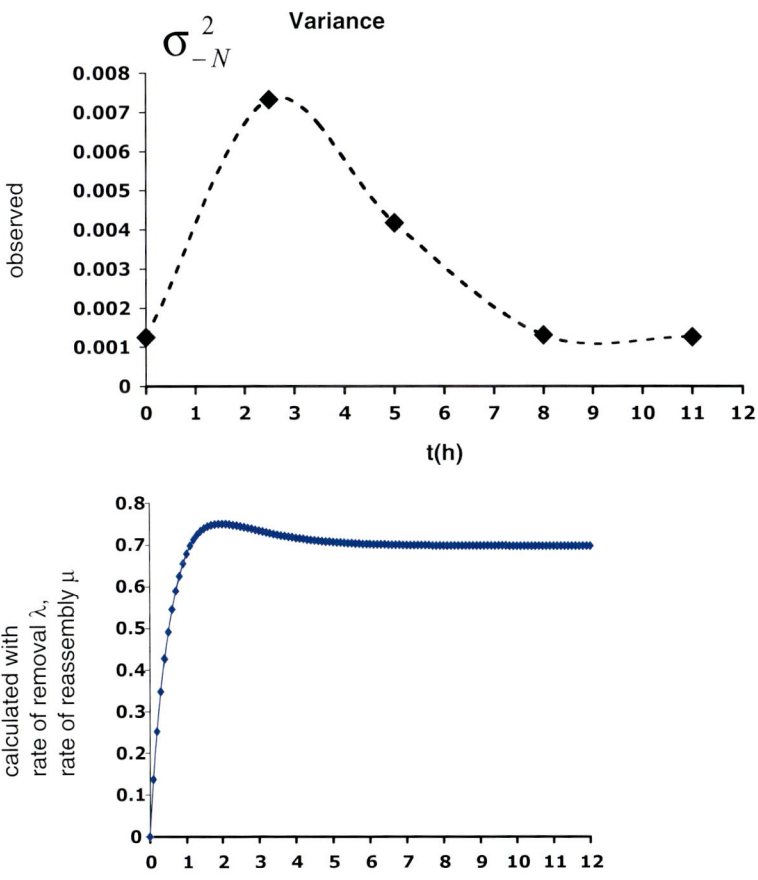

Figure 17. Statistics of *PHO5* promoter remodeling.

From further calculations, it emerges that the naked promoter state, devoid of all nucleosomes, is inconsistent with the data. It is disallowed. The activated promoter always retains at least one nucleosome. The slightly larger value of 1.1 that we measure reflects the ongoing process of nucleosome reassembly, competing with disassembly, which, in turn, accounts for the non-zero variance at late times.

With the omission of the naked state, we can easily account for all the experimental data (Figure 18). Specifying three rate constants, for removal of the first and second nucleosomes and for reassembly, suffices to account

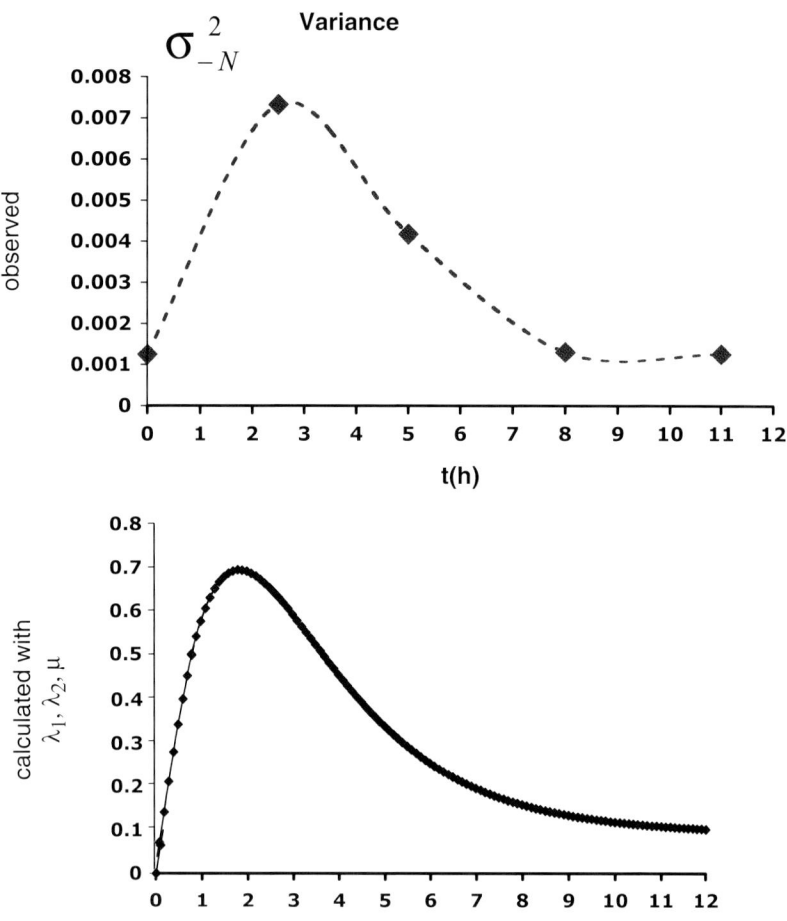

Figure 18. Statistics of *PHO5* promoter remodeling.

not only for the observed profile of the variance but also for the observed final distribution of nucleosomes, with the removal of 0.4, 0.82 and 0.68 nucleosomes, respectively, from positions N1, N2 and N3.

To summarize, the statistical analysis reveals two principles of chromatin remodeling. First, to reiterate, nucleosomes are not only removed but also reassembled upon transcriptional activation. Second, remarkably, the removal of nucleosomes from promoters conserves a single nucleosome. The value of 1 nucleosome per activated promoter is not an average over a broad distribution, but rather a property of every individual promoter. The remaining nucleosome is then statistically distributed over all promoter positions.

What possible mechanism might account for the removal of all but one nucleosome? The answer is suggested by the structures of RSC and RSC-nucleosome complexes, as determined by cryo-electron microscopy (Figure 19). RSC contains a large central cavity, and the RSC-nucleosome complex has extra density in the cavity. In this slab of the RSC-nucleosome complex, you see a close fit of a nucleosome core to the extra density in the cavity. The X-ray structure of the nucleosome could be docked to the RSC-nucleosome complex (Figure 20). RSC is seen to bind the nucleosome in a pincer grip, contacting the top and bottom surfaces of the nucleosome, while leaving the DNA exposed for remodeling. In a view from the top, with the upper portion of RSC removed, a single protein contact to the DNA,

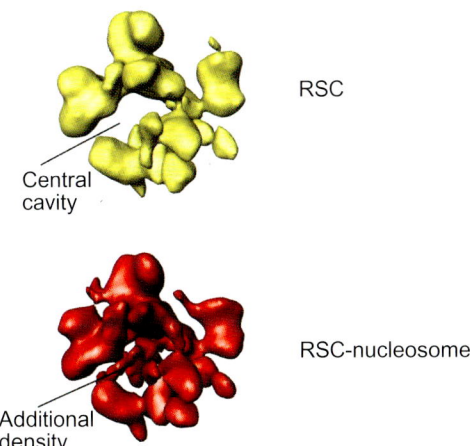

Figure 19. Cryo-EM structure.

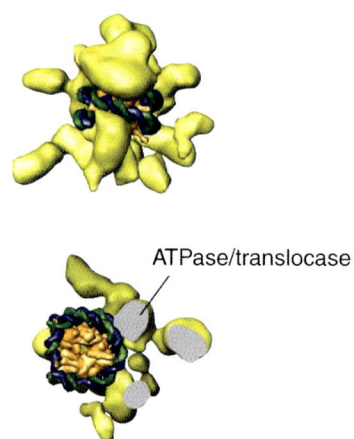

Figure 20. Structure of RSC-nucleosome complex.

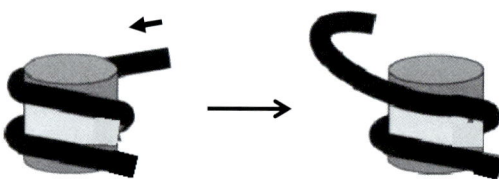

Figure 21. RSC-nucleosome complex invades an adjacent particle and unravels DNA.

due to the ATPase/translocase subunit, is revealed. It has been shown that RSC draws DNA in from one side and expels it from the other, resulting in sliding but not disruption of the nucleosome. The key to disruption is when the RSC-nucleosome complex slides up to a neighboring nucleosome. The complex will slide right through the neighboring particle, due to the equilibrium dissociation of DNA from the ends of the particle (Figure 21). DNA is unraveled and the histone octamer displaced from the neighboring particle as the RSC-nucleosome complex slides through. After transit of the complex through the promoter region, only one nucleosome, that bound to RSC, remains.

Finally, the chromatin structural noise decreases with time following activation. This behavior parallels that found years ago for the *PHO5* gene expression noise by a two-promoter method developed by others in bacteria

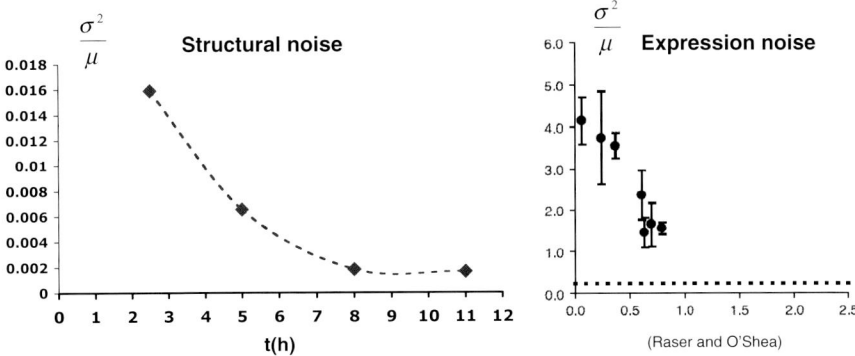

Figure 22. Nucleosome disassembly at *PHO5* is rate-limiting for transcription.

(Figure 22). We conclude that chromatin remodeling is rate-limiting for *PHO5* gene expression.

Prof. Roger Kornberg delivered the Albert Einstein Memorial Lecture in 2009.

References

Kornberg R.D. and Lorch Y. (1999) 'Twenty-five years of the nucleosome, fundamental particle of the eukaryote chromosome,' *Cell* 98:285–294.

Boeger H., Bushnell D.A., Davis R., Griesenbeck J., Lorch Y., Strattan J.S., Westover K.D. and Kornberg R.D. (2005) 'Structural basis of eukaryotic gene transcription,' *FEBS Lett.* 579:899–903.

Boeger H., Griesenbeck J., and Kornberg R.D. (2008) 'Nucleosome retention and the stochastic nature of promoter chromatin remodeling for transcription,' *Cell* 133:716–726.

Chaban Y., Ezeokonkwo C., Chung W.-H., Kornberg R.D., Maier-Davis B., Lorch Y. and Asturias F.J. (2008) 'Structure of a RSC-nucleosome complex and implications for chromatin remodeling,' *Nat. Struct. Mol. Biol.* 15:1272–1277.

Raser J.M. and O'Shea E.K. (2004) 'Control of stochasticity in eukaryotic gene expression,' *Science* 304:1811–1814.

Energy, Environment, and the Responsibility of Scientists

Yuan T. Lee

It is a great honor to be invited by the Israel Academy of Sciences and Humanities to present the Einstein Memorial Lecture. Two years ago, the activities celebrating the hundredth anniversary of the publication of three (or five) important papers by Einstein included a program called 'Physics Enlightens the World.' Scientific communities, with the help of business, organized in a big way to light up important parts of Taiwan, from the north to the southern tip of the island. Our program, which involved fireworks and a display of the $E = mc^2$ equation on Taipei 101, the tallest building in the world, was rated the best internationally.

When I started to carry out research on crossed molecular beams at the University of Chicago in 1968, we encountered Einstein almost immediately, in the form of the Bose–Einstein statistics. As we attempted to derive the interaction potential between two neon atoms from the measurements of differential scattering cross sections, we saw the oscillations in the angular distributions that are due to the 'even parity' of two Bosons. We were all excited to be observing the consequences of the Bose–Einstein statistics.

In what follows, I will not discuss Einstein's life or his science, but I believe I shall address some of his overarching concerns, as expressed by Gerald Holton in his article, 'Einstein and the Cultural Roots of Modern Science':

> When death approached to claim Einstein in April of 1955, his last acts were still fully in character. He remained strong-willed to the end, obstinately adhering to his ways. He had recently signed a manifesto with Bertrand Russell and others, intending to bring together the international community of scientists as a unifying counterweight against the divisive, national ambitions then rampant during the arms race.[1]

[1]Gerald Holton, 'Einstein and the Cultural Roots of Modern Science,' *Daedalus*, 27 (Winter 1998), pp. 1–44.

In the spirit of trying to 'bring together the international community of scientists as a unifying counterweight against divisive, national ambitions,' I wish to devote my lecture to the topic of 'Energy, Environment and the Responsibility of Scientists.'

1. A Historical Perspective on the Development of Human Society on Earth

Approximately ten billion years passed from the Big Bang to the creation of the solar system, about four and a half billion years ago. Four to five billion years from now, the diameter of the sun will expand to reach the orbit of the earth, and the earth may then cease to exist. The phenomenon of life on earth began about 4.4 billion years ago, and from then on, by means of photosynthesis and the formation of the biosphere, the ability of the earth's surface to capture and store energy from solar radiation changed appreciably. A couple of million years ago, after a very long period of evolution, humankind appeared on the heavily forested planet. At that time, the development of humankind and of human society was, on the whole, in harmony with nature. Humankind was indeed a part of nature, reliant on the sun to fuel the production of most of what it needs to survive. Since the human population was small, its limited activities seemed, for a long time, to have no great effect upon the biosphere or its own living environment.

However, the social development of humankind took a new turn after the beginning of the industrial revolution, about 250 years ago. Humans learned to transform energy from one form to another — from chemical, thermal and electrical to mechanical energy — and invented machines that could perform work thousands of times more powerfully, more precisely and more reliably than could be possibly done using the physical labor of humans or animals. The productivity of humankind increased immensely, and an unprecedented improvement in living standards was achieved. But during this process, large parts of humanity became addicted to using huge amounts of energy and began to depend more and more upon fossil fuels — coal and petroleum — which are buried underground and take millions of years to accumulate. To provide the energy needed for the production of various new materials, such as plastics, fertilizer, synthetic fibers, steel, concrete and cement, we have drastically changed the intimate relationship between humanity and nature. The harmonious relationship between humanity and the biosphere has been disrupted, and the important role played by the sun in human development seems somehow to have been forgotten.

As we enter the twenty-first century, we have begun to realize that the current development patterns of human society are not sustainable. Problems related to the population explosion, natural resource depletion and damage done to the living environment have grown increasingly serious. From a human perspective, the earth used to be regarded as infinite or unlimited, not only in terms of its available resources, but also in its ability to digest society's waste products. However, from the point of view of the damage already done to the ecosystem — the living environment — the earth must now be considered overdeveloped. Not only the developed countries, but also the so-called developing countries are overdeveloped. Unfortunately, the latter are following in the footsteps of the developed countries down the unsustainable path charted when the earth was still seen as unlimited.

Humanity must wake up, immediately, and recognize that human society is living beyond its means. We must learn to work together to find new, sustainable ways to re-establish an intimate relation with the biosphere, to live in harmony with nature, and to return to a more direct relationship with the mighty power of the sun.

2. The Dilemma of Living in a Half-Globalized World

The process of the globalization of human society that has occurred during the last few decades is only half complete. On the one hand, highly developed transportation and communication technologies have made our world seem smaller. The concept of the global village is taking root in a number of realms, most notably in the economic sphere. An example of the problems this has raised is the spread of disease: The thousands of airplanes that daily cross oceans and continents, loaded with people and goods, make it impossible to confine disease-causing bacteria, viruses and other microbes to specific locations. Similarly, it is increasingly evident that environmental problems, such as the depletion of the ozone layer caused by the use of chlorofluorocarbons and the global warming trends fueled by greenhouse gases, must be addressed on a global scale.

On the other hand, despite increased international collaboration in the areas of science and technology, high-tech-based economic competition is still largely carried out on a national basis. Currently, in the partially globalized world, only those able to stage their activities on a global scale are reaping the enormous benefits of globalization. It should not be surprising, then, that we face such problems as a widening gap between rich and poor — both between countries and between the inhabitants of individual countries — or that threats to solve these problems by military force have not disappeared. These problems might be avoided if the entire world were to become one community.

3. The 'Science and Technology in Society' Forum, Kyoto

In the fall of 2004, Mr. Koji Omi, Japan's former Minister of Science and Technology, organized the Kyoto forum on 'Science and Technology in Society.' More than five hundred leading scientists, business leaders and policy-makers from all over the world were invited to attend. The forum was so successful that it has since become an annual event.

At the first Kyoto forum's opening ceremony, Omi made two important points. First, the rapid progress of science and technology has both positive and negative aspects. The benefits of science and technology have not yet reached everyone equally, and this, as he said, 'is really what symbolizes the lights and shadows of science and technology.' The negative aspects must be properly controlled, while the positive features should be promoted.

Omi's other important point was:

> Today's problems are global and cannot be solved by any single country or by scientists alone. ... Boundaries between nations are merely lines on a map; nature makes no such distinctions. We should think of ourselves as members of mankind, whose very existence will be at risk if we do not live in accordance with the principles of Mother Nature.

I believe most of us would support this idea without hesitation, but its implications raise some complex questions relating to the limited nature of the earth and the world's 'half globalized' state — questions that we must address if we do not want our efforts to find solutions to lead to further problems. For example, we must ask, 'How many people could the planet support, if we were to extend the living standard of people in the developed countries to everyone on earth?' When India became independent, Mahatma Gandhi was asked how the people of that country could catch up with British living standards, Gandhi rightly responded that it would take the natural resources of many planet Earths to support a British way of life for all the people in India.

We need to appreciate fully the consequences of practicing the 'beneficial' sciences in a globalized, market-driven economy. Used mainly as a tool of global economic competition, those sciences, for all their benefits, can produce miserable losers.

4. Science Should Tackle Urgent Problems

The scientific community as a whole should take responsibility for ensuring that science and technology will benefit everyone equally. Otherwise, along with the material comforts and improvements in healthcare produced by our science, the population explosion and the excessive exploitation of natural resources that it generates stand to overload the planet. Eventually, our combined efforts can make sustainable development possible.

One of the most urgent problems faced by humanity today is that of the relationship between energy and the environment, in particular the global warming trends caused by the emission of greenhouse gases, and the energy crises brought on by the widening gap between the limited supply of petroleum and the rapidly growing demand for it. A second problem that menaces large portions of humanity is the spread of infectious diseases, like those caused by the H_5N_1 (bird flu) virus.

It is comforting to know that, at present, the amount of solar energy received by the earth's surface in one hour is approximately equal to the world's total annual energy consumption. In other words, the amount of received energy is approximately ten thousand times that consumed by human society. If an inexpensive, practical photovoltaic cell that could convert 10% of the solar energy it received to electricity were to become available, it would take only 1% of the planet's land area to generate enough electrical energy to satisfy the energy needs of the entire world. If the electrical energy generated by a photovoltaic cell could be used to electrolyze water into hydrogen and oxygen — or if water could be dissociated even more directly by using a combination of photovoltaic cells — it is not inconceivable that countries with large land masses could become energy exporters, or that hydrogen gas could become a major energy source, initiating an age of 'hydrogen economy.' Alternatively, if we learn to develop biofuel more efficiently, photosynthesis could be used to satisfy the need for liquid fuel now met by petroleum.

To make it possible for the world to achieve sustainable development, we need to take the following steps to reduce our dependence on fossil fuel:

(a) We must increase our energy efficiency.
(b) We must develop efficient renewable energy sources, such as photo-voltaic cells; wind power generators; methods of utilizing geothermal energy, ocean flow and temperature differential; and biofuels of various kinds.
(c) We must develop a new generation of safe nuclear reactors (along with appropriate waste disposal techniques) and push the development of fusion reactors.
(e) We must examine our population policies and lifestyles.
(f) We must protect our living environment and ecosystems and maintain biodiversity.

Current scientific knowledge and technology enable us to take initiatives in all these directions, but many challenging problems await solutions. For example, in photosynthetic processes, most of the solar energy is stored in plant fibers rather than in carbohydrates. Production of alcohol from sugar cane and corn has been effective and successful; however, the challenge lies in the effective production of alcohol from fiber through hydrolysis and fermentation. For harvesting geothermal, ocean-flow and temperature-differential energy potentials, new engineering technologies need to be developed.

I am quite optimistic. Forty to fifty years from now, I believe, we will largely be free of the use of fossil fuels. We will again be directly reliant on the powerful sun, perhaps with the assistance of 'micro-suns' in the form of fusion reactors.

But during the transition period of the next 30 years, especially until the fusion reactor becomes a reality, we probably will continue to depend to a great extent on coal, though nuclear fission reactors will also play a role. The sequestering of CO_2 will remain a problem in need of a solution.

It is also comforting to know that the long-neglected development of new vaccines for infectious disease is finally picking up, in the wake of international efforts. The race is on between the perfection of an H_5N_1 vaccine and the mutation of the H_5N_1 virus that would make it transmissible among humans. More research needs to be carried out in this area. However, we need to be aware that the funds spent globally on medical research have been targeted to problems relating to 10% of the world's population. If we fail to pay attention to the deteriorating situation in developing countries, there is no way we will be able to combat infectious disease effectively.

5. Sharing of Scientific Knowledge and Technologies in a Globalized World

For centuries, the scientific knowledge accumulated by humankind has been shared among scientists quite freely. Scientists generally still follow Francis Bacon in believing that the knowledge accumulated through their efforts should be shared by all. Early in the last century, Madame Curie was asked why she didn't apply for patents to her discoveries; had she done so, she would have become as wealthy as Thomas Edison. She replied, simply, that she did not wish to take that advantage, because she believed that scientific knowledge should belong to all of humankind. It was her idealism that inspired me to become a scientist. In modern society, however, when scientific knowledge is developed, transformed into technology and put to use in society, it becomes a basis for economic competition. Protection of patents and intellectual property rights has become a critical issue. Knowledge sharing now stops at basic research and 'pre-competitive' technology; 'competitive' technology is excluded. However, the gap or time lag between scientific discovery and technology in the marketplace has become shorter and shorter. It was a century for automobiles, five years for computers and only eighteen months for microprocessors. Now, in certain areas of scientific investigation, it is no longer possible to distinguish between basic research and associated competitive technology.

As the relationship between science and technology has become closer, the dilemma of 'to share or not to share' has become an important issue — not only for the application of technologies, but also for the basic scientific discoveries themselves. It certainly does not seem fair for some countries to produce most of the scientific knowledge in the public domain, while others dedicate themselves mainly to protected, mission-oriented technological development in order to gain economic competitiveness. In a market-driven economy, free and open economic competition and adequate protection of intellectual property rights are necessary for development. Yet, we must ask whether, in a highly globalized world, we can find a new and better way to allow both the creation and the sharing of knowledge and technology to take place in a more orderly fashion, for the sake of promoting sustainable development for the entire world. Strong global public support for the advancement of science and the development of technology, and for shortening the patent protection period, might move us along in that direction.

In recent years, in the fields of high-energy physics and astronomy, scientists share their knowledge quite freely and have been more willing to help each other across national boundaries. In the field of biology, on the other hand, scientists tend to protect their intellectual property rights more tightly. At international conferences on biological sciences, scientists are often seen trying to learn as much as possible while revealing as little as they can on critical issues. Whether this may be attributable to the fact that high-energy physics and astronomy are supported by public funds, while certain areas of biological research are dominated by for-profit pharmaceutical industries, is worth studying in detail.

Many of the problems we face today cannot be solved using current scientific knowledge and technologies; they await the accumulation of new knowledge and the development of new technologies. That is why it is so important for us to continue our efforts to advance science and technology, and to educate a new generation of creative scientists. On the other hand, scientists must realize that science and technology cannot solve all the problems we face. On the contrary, the rapid development of human activities and especially the fast-moving global economy, propelled by the advancement of science and technology, may create further problems as contact among peoples becomes more intimate.

Although the globalization of the world economy is driving us toward a borderless society, it will not reduce the differences among peoples overnight. Establishment of a common global culture, together with more effective ways of communication among the world's peoples, will take time. The differences among cultural heritages, languages and religions that make this world so rich and colorful will not and should not be made to disappear. Whether or not the differences between civilizations must cause them, inevitably, to clash (as suggested by Samuel Huntington) would seem to depend upon how well people around the world learn to communicate and to understand, appreciate and respect cultural heritage. If we are to become good citizens of the global village, we need to learn quickly and to teach our young people to take a global view, and to respect, appreciate and understand cultural differences. In this regard, scientists can certainly lead the way.

6. Concluding Remarks

Over the long history of humankind, our ancestors invented various technologies in order to survive better or improve their quality of life. Their curiosity and desire to understand natural phenomena were the basis for the advancement of science. One hundred years ago, the advancement of science was still driven by the available technology; only in recent years have technological advances been driven by the results of scientific research.

At the 2006 'Science and Technology in Society' Forum in Kyoto, one of the participants touched upon the relationship between research, development, innovation and money. As he put it, research and development are responsible for transforming money into knowledge, while innovation turns knowledge into money. This description may be oversimplified, but it is surely true that without generous investment in research and development, curiosity-driven scientists will be able neither to generate new scientific knowledge nor to develop new technologies, and many urgent problems will remain unsolved.

In recent years, encouraging improvements have occurred in international scientific collaboration. Many projects have been initiated and many agreements signed. Year after year, we have discussed building science, technology and education capacity in developing countries. The worsening situation of the world as a whole, however, has yet to reach a turning point. For example, rainforests, whose function in the biosphere is so often compared with that of the lung in the body, are continuing to disappear from the earth's surface. Every summer for the past decade, we have observed the thick, dark smog generated by forest fires in Indonesia, contaminating the air not only there but also in neighboring countries. It is not realistic to blame Indonesia for this, or to expect it to be able, on its own, to keep its rainforest from disappearing. Unless we ourselves take responsibility for the protection of that forest — for example, by raising funds to help Indonesia establish a protected 'global rainforest' — then, no matter how seriously we engage in international scientific collaboration, the rainforest will continue to disappear.

In order for science and technology to solve the problems humanity faces in the twenty-first century, it is not enough to advance science and technology at a faster pace, though such advancement certainly will continue shaping the development of human society. Unless we give our attention to the positive and negative roles played by science and technology in this finite and 'half-globalized' world; unless we learn to work together across national boundaries and push 'global competitiveness' in solving problems related to sustainable development, rather than worrying only about the 'national competitiveness' of our own countries — the problems will not be solved.

The best way to transcend national boundaries is to make them disappear altogether. Although it may take a long time, our future will depend upon how soon all of us in different countries learn to operate as a single world community. We do not have much time to waste. Perhaps, halfway through the twenty-first century, a 'United States of the Planet of Earth' or 'Global Union of the Planet of Earth' will begin to take shape, enabling the world's sustainable development. Otherwise, in the not too distant future, the solar system may send to earth a farewell message for humankind.

Prof. Yuan T. Lee delivered the Albert Einstein Memorial Lecture in 2007.

Res Ipsa Loquitur: History and Mimesis

John E. Wansbrough

When in London Albert Einstein, following upon the Royal Society's successful expedition to photograph a solar eclipse, described the movement of bodies as contingent on a 'system of co-ordinates,'[1] he observed an ancient and general principle in the organization of all experience: namely, that empirical data were of use only insofar as they could be related to a field of perception already plotted. The principle was of course analogy, and the system of co-ordinates an essential framework for making what is strange and unruly into the familiar and orderly: in another words, an exercise in intellectual domestication.

In these days of intense speculation on why and how we think what we think, analogy is so much taken for granted that all mystery must seem to be accounted for, and all data in jeopardy of becoming 'obvious' (to employ a current catchword) or self-evident. Hence my selection of title for this important occasion, in which I am hardly qualified to participate, but nonetheless sensitive to the great honor of your President's invitation to do so. Naturally, it is not all that difficult to discover some point of contact between my interests and those of Einstein, whose thought and activities comprehended most of the human condition (Terentius: *Homo sum, humani nil a me alienum puto*). My subject this evening is (and for a very long time has been) the nature of historical discourse and its apparently endless proliferation of literary expression.

Acknowledgment of historiography as literature is, though somewhat grudging, now fairly widespread. This may be nothing more than recoil from attempts to make of history a fully fledged science (recently dubbed 'Cliometrics'), but 'literature' here seems to imply little more than the fact that history is usually written in narrative prose, with, as one historian put it, 'the added constraint of factuality.'[2] On both counts, of narrative style and of factuality, the assertion may be thought just a little ingenuous, as

---

[1] In response to an invitation from the *Times*, 28 November 1919.
[2] Cf. J. Barzun, *Clio and the Doctors*, Chicago 1974, esp. pp. 54–59, 116–118.

any serious student of 'literature' is bound to observe. It is well known that Aristotle reckoned history among his literary genres, but with the significant observation that its proper domain is the particular descriptive statement from which nothing relevant might be omitted.[3] That was in contrast to poetry, characterized by the universal truth of a general statement. Two remarks seem pertinent: (1) Aristotle's distinction between the two genres turns upon the implicit (!) role of *referent* in historiography, about which he is somewhat naive; (2) his definitions are embedded in a discussion of *mimesis*, about which I will in due course have something more to say.

At least the scene was set, some twenty-five hundred years ago, for an analysis of what the historian ought to be about. In the view of Aristotle, it might seem, his task was to depict in the most minute detail the events of the past. There are some, even today, who suppose their task to be 'discovery of a pre-existing true state of affairs.'[4] Most, however, recognize that they must settle for something less than that, namely, a selectivity that in turn not merely imposes upon them choice of topic but also a corresponding stylistic constraint. What exactly, in other words, is a sentence?

With that question I am admittedly compelled to trespass upon the domains of the linguist, the philosopher and the literary critic, as well, of course, as that of the novelist. In a typically provocative essay, Clifford Geertz has shown how the traditional disciplinary lines of demarcation have been dissolved. In an impressive parade of names from Steiner and Lévi-Strauss to Doctorow, Borges and Nabokov, the intentional blurring of genres is demonstrated, to an extent that must obliterate the ancient and time-honored distinction between history and fiction.[5] Now, whatever one might think of this development, it is clearly here to stay and must cause some unease among historians who had staked a claim on their special ability to tell us 'what really happened' (*wie es eigentlich gewesen*). Thus, the 'language game' has got to be played, and, moreover, from the premise that *text* is the primary datum of human experience. Further requisites are a literate public, a concept of 'reading' as productive, and curiosity about

[3] *Poetics*, 1451 and 1459; cf. G. Grube, *The Greek and Roman Critics*, London 1965, pp. 83–85; R. Humphreys, 'The Historian, His Documents, and the Elementary Modes of Historical Thought,' *History and Theory*, XIX (1980), pp. 1–20.
[4] E.H. Carr, *What is History?*, London 1961, esp. pp. 85–102; cf. J. Price, review of Carr, in *History and Theory*, III (1964), pp. 136–145.
[5] C. Geertz, 'Blurred Genres — The Refiguration of Social Thought,' *The American Scholar*, XLIX (1979–1980), pp. 165–179.

'writing' as not merely interpretative but creative in the ontological sense (*Gestaltung*).

One consequence will be the need for historians to explain, in *post*-Aristotelian terms (!), how what they do is different from writing novels. Both those ancient parameters, 'referent' and 'mimesis,' will undoubtedly benefit from further scrutiny. It is no longer enough to be assured that the 'sources' tell us this or that. The very prose in which the assurance is expressed has become suspect. To adduce many instances would not be so difficult but certainly distracting. What I propose here is examination of two such, actually quite dissimilar, but which for methodological reasons appear to have attracted a very similar if not quite identical treatment.

My first example is the commentary generated by a *significant* (!) portion of Arabia in the seventh century CE. Of that there exists a good deal, in a more or less continuous stream from then until now. 'Stream' is perhaps not the right word: 'torrent' might be more appropriate to the volume of literature provoked by the uninterrupted effort to depict the origins of Islam. And that is the first point I should like to make. The very quantity of the corpus must figure in its critical assessment. Like the Mongol conquest, the discovery of the New World and the French Revolution, that remote Arabic 'event' now constitutes a major preoccupation of the historians' guild. From this position it has undoubtedly profited, and that is my second point. Quantity not merely produces but determines quality: Very few are the exegetical methods that have not been, in the course of this long and arduous confrontation with the past, exploited in an attempt to understand that literature. I see these as falling into one of the three following categories: (a) Islam as the re-casting of pre-Islamic Arabia; (b) Islam as the product of minority (external) historiography; (c) Islam as the response to interconfessional (Judaeo-Christian) polemic, what I have elsewhere essayed to describe as the 'sectarian milieu.'[6] Now, it is not my intention this evening to burden you with what I have offered historians by way of exegesis. The passage of time involves a significant intellectual therapy, and while I have not moved in any of the directions so ardently advocated by my many critics, I have managed to move, and this must be a token of some residual vitality. The process is standard and thus familiar: Comparison throws up as many antitheses as it does analogies, and it was by juxtaposition of this first example with my second — a *not so significant* (!) portion of Syria

---

[6]J.E. Wansbrough, *The Sectarian Milieu — Content and Composition of Islamic Salvation History*, Oxford 1978, passim, but esp. pp. 32–49 and 114–119.

in the fourteenth century BCE — that I was impressed by the enduring obstinacy of historical method.

But let us consider for a moment that remote portion of Arabia. Bereft of archaeological witness and hardly attested in pre-Islamic Arabic or external sources, the seventh-century Hijaz owes its historiographical existence almost entirely to the creative endeavor of Muslim and Orientalist scholarship. Though I am obliged to add that these have seldom been found in collusion, there is an impressive unanimity in their assent to the historical 'fact.' Since the evidence, or its absence, is common to both traditions, it might be thought that they share certain methodological presuppositions. These could be set out in the following ways:

(a) as 'paradigm' = the general hypothesis according to which empirical and other data are perceived;[7]
(b) as 'structure' = the system of coordinates by which analogy and internal consistency are established;[8]
(c) as 'linguistic closure' = the syntactic and semantic constraints imposed by selection of a vocabulary to depict events in language.[9]

Now, together these rubrics are meant to comprehend the sum of techniques available to the historian in his exegetical task. They also happen to describe the means available to any writer, or, for that matter, speaker, whose intention it is to convey an impression (and it can be no more than that) of his own or someone else's experience. Appeal to 'common sense' is merely recourse to a (one hopes) shared paradigm; explanation can only be insistence upon an intelligible choice of structure; and style must inevitably reveal a personal decision about the adequacy of language to the task of description. This is not to say that historical or any other literature can persist (at least for very long) in a condition of solipsism. All expression is constrained — indeed imprisoned — by the grammar of a sentence.

And what has all that got to do with the seventh-century Hijaz? I would say approximately this: The sources for that historical event are exclusively literary, predominantly exegetical, and incarcerated in a grammar designed

---

[7]S.C. Pepper, *World Hypotheses — A Study in Evidence*, Berkeley–Los Angeles 1966, esp. pp. 115–137; T.S. Kuhn, *The Structure of Scientific Revolutions*, Chicago 1970, esp. pp. 43–51.

[8]R. Barthes, 'Historical Discourse,' in *Structuralism — A Reader*, London 1970, pp. 145–155; J. Culler, *Structuralist Poetics*, London 1975, passim, but esp. pp. 96–109.

[9]Wansbrough (above, n. 6), pp. 141–142 ad M. Arkoun, 'Logocentrisme et vérité religieuse dans la pensée islamique,' *Studio Islamica*, XXXV (1972), pp. 5–51.

to stress the immediate equivalence of word and world. Or, I might be inclined to add: All we know is what we have been told. With neither artifact nor archive, the student of Islamic origins could quite easily become victim of a literary and linguistic conspiracy. He is, of course, mostly convinced that he is not. Reason for that must be confidence in his ability to extrapolate from the literary version(s) what is likely to have happened. The confidence is certainly manifest; the methodological premises that ought to support, or, at least, accompany it, are less so. One can only suspect the existence somewhere of a tacitly shared paradigm, that is, an assumption that the literature in question has documentary value. Such it has, indeed, though not quite in the sense here supposed. However that may be, the assumption itself might seem to be corroborated by a further curious circumstance: I mean the near absence of Islamic data from comparative studies of religion. The material from which relevant data could so easily be culled has come to be regarded as sui generis, as though of value only for the unique historical phenomenon it purports to depict. Now, while all historical phenomena are admittedly unique, the means of describing them are severely limited. I refer to linguistic constraints: Whether these entail, or merely reflect, conceptual ones, is a problem I am unable to solve. In any case, the constraints themselves permit erection of a 'system of co-ordinates,' and thus discovery of the analogies indispensable to description. Of course the procedure can be exaggerated, and we have had warnings enough about the dangers of 'parallelomania,' at least when defined as historical diffusion. But that definition is neither complete, nor, for that matter, necessary.

Reading literature as history is a common if controversial pastime. While I am often tempted to respond by reading history as literature (and have frequently been accused of this impropriety), there is surely some more practical mode for making the transfer from unique event to general proposition. In order to deal with the reports of seventh-century Arabia, I divided the field into constants and variables: the former representing the 'basic categories' common to most descriptions of monotheism; the latter representing 'local components' that give each version its special character. Recourse to this simple taxonomy seemed to facilitate a discussion of Islamic origins in terms that would make sense to any student of religion, in short, to make of the unfamiliar an intelligible unit of study. The constants were prophet, scripture and sacred language; the variables were the specifically Arabian features of these, together with such traces of local usage as could be inferred from its later abrogation by the new faith (e.g., in ritual practice and civil law). In this scheme of things, the problem of diffusion need not, but inevitably does, arise. The obvious, and certainly easier, alternative is to calculate the factor of polygenesis: that is, prophets are the agents of divine revelation which, once recorded, must contain the sacred language of God's word. That calculus does not, of course, yield a specifically Mosaic exemplar for Muhammad or a Davidic genealogy for Jesus. But those are variables and of only marginal interest to the structural study of religious phenomena. However, my proposals have found favor with neither Muslims nor Orientalists, and that, I suspect, for the very reason that historians regard their task as the elucidation not of constants but of variables. The paradigm, I have suggested, is Aristotelian and just possibly in need of revision. The two points at which such might be undertaken are the concepts of referent and mimesis.

Though mostly employed as an existential, hence empirical, concept, 'referent' requires for all but the contemporary chronicler an act of faith. The act is of course not quite arbitrary: it is sanctioned by guild membership. One reads the works of one's colleagues, and, sooner or later, something like a consensus emerges. In most cases that will have been underpinned by the 'hard,' if often mute and impenetrable, evidence of archaeology and/or archive. But not always. And I have referred repeatedly to the literary and exegetical character of the sources for the seventh-century Hijaz. The implicit caveat is heard but seldom heeded. The notion of literary 'convention' must be in some way found abhorrent, for there is a perennial urge to substitute for it historical 'reality.' (I am here reminded of the recurrent question asked by my children, when many, many years ago I used to read aloud to them in that last hour before bedtime such classics as *Oliver Twist* and *David Copperfield*: 'Is it true?' they would ask. They meant of course 'Did it happen?' and while I could hardly assure them of that, I was able to say that it was very true indeed.) In other words, 'referent' may also function as a literary convention, as that attractive (because reassuring) link between experience as reader and experience in life. But if 'referent' is a psychological necessity, its historicity is not thereby confirmed. Now, Aristotle told us that the purpose of history, as a literary genre (!), was to relate the event in all (not merely its significant!) detail. He took into account neither the fallibility of the eyewitness nor the constraints of the medium (=language) available to him for that task. I have already intimated that his assessment was ingenuous. Unfortunately, he uttered no further word on that particular subject, and one must suppose that those otherwise precious powers of analysis were in this instance satisfied with what everyone knows to be 'common sense.'

In respect of mimesis Aristotle had rather more to say. Much of it is widely familiar as an analysis of representation, and in particular of mimicry and imitation.[10] The context is tragedy and the examples theatrical. Of epic mimesis he thought rather less, and found only Homer to be an unqualified success. It was this treatment of the subject, defined as 'the reproductions of reality,' that generated the now classical monograph of Erich Auerbach.[11] But elsewhere, Aristotle employed the term 'mimesis' to describe the relation of numbers to geometric figures, and thus introduced, as it were, a new dimension into the argument.[12] That was the condition that the mimetic process involved transfer to a different medium, a postulate not so easily derived from his analysis of tragedy. Epic poetry might have provoked this *aperçu*, as would have history, but in the event did not. It is of course the notion of a new (or different) medium that requires a definition of mimesis not as 'reproduction' but as 'production of reality.' And that provides a rationale for the creative licence to which so few historians are inclined to lay claim. Grounds for this modesty must be manifold, and I would not dream of trying to identify them. What must, however, be said is that historiography, like every other kind of literature, does employ a new medium. That medium is language, which involves willy-nilly its own set of constraints. For example, nothing can be linguistically depicted except as linear and sequential. That very meaning of 'syntax' = order generates in narrative prose a capsular consistency that, in the context of historical discourse, takes on an uncanny resemblance to logic and causality. Language is also constrained by semantic association: Every unit evokes not merely itself, but also its antithesis and a penumbra of metaphorical and metonymical reference. Employment of such simple and apparently unambiguous epithets as 'regalian,' 'sacral,' 'urban,' 'mercantile,' etc. must entail for every reader and, more important, every writer a concatenation of acquired imagery that can hardly be presupposed or, more important, pre-controlled.

Now, my purpose in adducing these homespun truths is to remind you of this simple and quite straightforward precept: the historical record consists of nothing more or less than human utterance and ought to be assessed by reference to all the criteria now assembled for this very rewarding task, If

---

[10] *Poetics*, 1447–1462.

[11] E. Auerbach, *Mimesis — Dargestellte Wirklichkeit in der abendländischen Literatur*, Bern 1946, passim, but e.g., p. 183 ad Dante.

[12] *Metaphysics*, V, 14; cf. V. Zuckerkandl, 'Mimesis,' *Merkur*, XII (1958), pp. 225–240; J.E. Wansbrough, *Bulletin of the School of Oriental and African Studies*, XXXIX (1976), pp. 443–445.

I have managed (and this is all but certain) to persuade you that what we know of the seventh-century Hijaz is the product of intense literary activity, then that record has got to be interpreted in accordance with what we know of literary criticism. My own experiment, in terms of structural features and formulaic phraseology, was never intended to be more than that: an experiment. Reactions to it provoke the impression that to historians the factor of ambiguity is not especially welcome. What seems to be required is some kind of certainty that what is alleged to have happened actually did. I doubt very much whether, for this particular segment of the story, we can attain that certainty: the requisite material is not to hand. And that is the purpose of my second example of historical mimesis. Here, scholarship basks in an almost unique condition of liberty: the sources are exclusively archaeological and the record innocent of any contextual analogy to standard models. That these basic conditions have not deterred historians from erecting a 'system of co-ordinates' and from 'discovery of a pre-existing true state of affairs' must tell us something about the dedication to 'fact' of that professional guild.

About the second example: The phrase 'a *not so significant* portion of Syria in the fourteenth century BCE' is merely intended to convey the absence of an exegetical factor in the extant record from the Bronze Age settlement at Ras Shamra known to us as Ugarit. Its traces are severely and literally 'objective': These include a remarkably heterogeneous range of artifacts, several collections of cuneiform tablets exhibiting at least six languages, evidence of municipal, religious, funerary and domestic architecture, distributed in an urban plan containing carefully executed portions of enclosed and open space, on a site so far estimated to be an area of fifty acres, to which may be added the nearly adjacent coastal sites of Minet el-Beida and Ras Ibn Hani. Even without external support, all this had to be capable of yielding some sort of image for a toponym virtually unknown to Orientalist scholarship until the discovery of the Amarna correspondence.[13] Circumstantial evidence, subsequently perceived, has been only marginally helpful: e.g., random attestation in cuneiform (Elba, Mari, Alalakh, Palestine) and Egyptian (Karnak, Memphis) sources. The full

---

[13]I.e., 1887: J.A. Knudtzon, *Die El-Amarna Tafeln*, Leipzig 1906–1915, esp. pp. 308–318 and (O. Weber) 1016–1017, 1097–1102; cf. C. Kühne, *Die Chronologie der internationalen Korrespondenz von El-Amarna* (Alter Orient und Altes Testament), Neukirchen-Vluyn 1973; J.G. Heintz, *Index documentaire des texts d'El-Amarna*, Wiesbaden 1982.

chronology of Ugarit is almost entirely notional: 'fourteenth century BCE,' based on Amarna, is symbolic of a possible millennium 2200–1200.

'Significance,' in other words, has had to be read into, not out of, the traces. The process might be described as one of metamorphosis: from discrete and antiquarian remnants toward a legible pattern of meaningful experience. That this could be achieved at all required a good deal of imagination and the application of several techniques in essence and fact quite different from those of the literary critic. Here we have no commentary for analysis, which is all we had for the seventh-century Hijaz, but rather, an abundance of hard and mute 'fact.'

So confronted, the historian of the ancient Near East has been compelled to adopt at least one — often more — of a number of strategies for expression of these data. In theory his choice might appear to be unlimited; in practice it has been unexpectedly restricted. Reason for this must lie somewhere in the acceptance of a paradigm for assessing discrete and random witness (archaeology is after all notoriously unpredictable): i.e., it can only be read in terms of a pre-figured system of coordinates. Selection of the system will in turn depend on what is already available. For Ugarit the choice comprehended several (vaguely) contemporary models (themselves hardly certain in their political and socio-economic contours): e.g., Hittite, Aegean, Cypriot, Canaanite and Egyptian. The manner in which Ugaritic data have been slotted into these unstable structures inspires only qualified confidence. For example, the site has been described both as a 'maritime metropolis' and as a 'territorial state.' One might be excused for supposing that it was none of the above-mentioned contemporary contexts, but rather Venice, that supplied this particular model. The maritime dimensions of Ugarit are traced from the Aegean via the coast of Hittite Asia Minor and Cyprus to Egypt (on the basis of some very ambiguous documentation); its territorial dimensions are estimated to include some sixty kilometers of coastline by about forty kilometers of hinterland = 2400 sq km of political hegemony, containing 195 named localities with a population of around 25,000.[14] Evidence for that reconstruction has been derived from the occurrence of toponyms in Ugaritic chancery records, none of which provides unequivocal witness to the political entity so depicted. But that was only the beginning. Once the general situation of Ugarit had been staked

---

[14] A panoply of this exegesis may be found apud M. Liverani, *Stoia di Ugarit nell'età degli archivi politici*, Rome 1962; M. Astour, 'Ugarit and the Great Powers,' in *Ugarit in Retrospect*, Winona Lake (Indiana) 1981, pp. 3–29; M. Heltzer, *The Internal Organization of the Kingdom of Ugarit*, Wiesbaden 1982 (with reference to earlier studies).

out in the interstices of surrounding archaeology, it seemed easy enough to fill the gaps by recourse to a series of case studies, each the product of a separate comparison with materials quite disparate in time and space.

Perhaps the most remarkable, and certainly the best known, have been those adduced to support a reconstruction of culture in Ugarit: Its language is described as Proto-West Semitic (mostly via Classical Arabic), its literature is deemed Canaanite epic (mostly via Biblical Hebrew) and its religion interpreted as a version of ancient Near Eastern mythology (via tenuous correspondence with the theophoric nomenclature of a Semitic pantheon).[15] While none of these postulates is entirely without substance, the first two might be thought to suffer from a kind of diachronic disability, and the third from a generous proportion of unaccounted for onomastic. At least two characteristics of the procedure inherent in this exercise are salient: (1) the easy metamorphosis of the philologist's hypothesis into the historian's 'fact'; and (2) the reconstruction of Ugarit as a source or vehicle of subsequent evolution. The methodological significance of both is enormous. Together they constitute the paradigm of historical explanation. One works, after all, from established fact toward a linear sequence of development. Nothing is more welcome than that which can be seen to herald the later circumstance, even or perhaps especially when its intrinsic ambiguity has been interpreted precisely to that end. The circularity of this logic has of course been noticed, but seldom taken fully into account in the actual calculation of results. To this day the Ugaritic language, even its alphabet, is something of an enigma, its literature only barely elucidated (and certainly not in a linear development that could have produced the Hebrew Bible), and its religious expression remains incarcerated in a plethora of as yet unexplained godnames and rituals.[16] But that is not to say that the reconstruction so far generated is without value. Every configuration of data has got to be of some use, if only to remind us of its methodological limits. But the sum of such lucubration is less important than the means by which it was delivered.

Less well developed, but gathering gradually in substance, are the 'case studies' concerned with the political entity called Ugarit. Here all available data have been assimilated to a model of monarchic authority: not merely monarchic, but autocratic in expression and dynastic in

---

[15]E.g., S. Segert, *A Basic Grammar of the Ugaritic Language*, Berkely–Los Angeles 1984; J.C.L. Gibson, *Canaanite Myths and Legends*, Edinburgh 1977; J.C. de Moor, 'The Semitic Pantheon of Ugarit,' *Ugaritforschungen*, II (1970), pp. 187–228.

[16]Cf. J.E. Wansbrough, 'Antonomasia — The Case for Semitic 'tm,' in *Figurative Language in the Ancient Near East*, London 1987.

transmission. Once adopted, this interpretation has dictated the course of further description, e.g.,

(a) internal administration: the 'king' as initiator and final arbiter of executive decisions;
(b) external relations: the 'king' as sole respondent in negotiation, whether in tributary or autonomous status;
(c) military organization: the 'king' as sole donor of rank and authority;
(d) naval organization: the 'king' as disposer of fleet movement and allocation;
(e) economic activity: the 'king' as source and exclusive principal of commercial transactions.

Evidence for this remarkable versatility has been found in chancery records, admittedly plentiful but also notably lacunal in their coverage of the transactional apparatus. But once linked to a familiar model, the gaps could be filled by resort to imaginative reconstruction. Like all such, the 'regalian' model exhibits an *a priori* decision about the relevance of archaeological/archival data. It is here not a matter of selectivity, but of a hermeneutic grid by means of which all the available material could be processed. The result was thus predetermined. The method is admittedly a standard one and hardly without precedent. Its point of departure, however, is nothing more than a reading of the West Semitic term, 'm.l.k.' as unambiguous reference to 'kingship,' a meaning it did eventually acquire, but rather later than the period in question. Without that gratuitously adduced ingredient, the chancery records of Ugarit attest to the indisputable activity of a merchant oligarchy exhibiting the normal gain-motivated behavior of businessmen.[17]

In this very context of source analysis a further point could be made. The polyglot chancery of Ugarit, to which I have already referred, has been traditionally aligned with the practice of contemporary and landlocked models served by Akkadian as *lingua franca*. While the abundance of tablets in that very language must attest to its widespread use, that can hardly be adduced as witness to its exclusive employment for international relations. One needs little more than the material pertinent to contact between Cyprus and Ugarit to suppose that in the Levantine context the Ugaritic language enjoyed intelligibility far beyond the confines of the metropolis. This surmise might also benefit from a historian's analogy: If the later

---

[17]Cf. J.E. Wansbrough, 'Ugarit — Bronze Age Hansa?' in Karl Reinhold Haellquist (ed.), *Asian Trade Routes*, London 1991.

Phoenician commercial expansion did not depend on, it almost certainly profited from, the concomitant spread of its local idiom. But even without this, one could guess from the Ugaritic finds beyond Ugarit that communication might occasionally take place outside the strictures of a complex and arduous school tradition (which is the only way that Akkadian can be described). Moreover, the respective distribution in the chancery records of Ugaritic and Akkadian scarcely shows demarcation along the lines of internal and external business — that is, Akkadian is abundantly exhibited in both spheres. It is tempting to suppose that selection is directly related to scribal training and a certain degree of experiment. The creation of a chancery rhetoric is the product of several variables, of which only the most obvious is communication. With a single exception, itself merely a paraphrase, we have no instance of a document in both Ugaritic and Akkadian versions. A provisional conclusion would have to be that the chancery scribe wrote the language he knew best. On the other hand, it would not be amiss to acknowledge the fragmentary character of archaeological data.[18]

In what, then, does the 'significance' of Ugarit consist? Its 'factuality' can be hardly be disputed; its meaning, however, is a methodological construct. While this ought to provoke no particular surprise, it may be worth mentioning a recent application of the data. In his study of economic structures in the ancient Near East, Morris Silver made liberal use of the Ugaritic material to demonstrate the existence of a market economy in the second millennium BCE. As must be well known, the argument is addressed to the thesis of Karl Polanyi which asserted the opposite, namely, that economic transactions in the Bronze Age were initiated and implemented from a regalian center, what is, in other words, professionally defined as a 'palace economy.'[19] While in my view Silver's interpretation of the data is emphatically sounder than Polanyi's, it must be said that both are economists, and thus dependent upon the exegesis made available by

---

[18]Cf. J.E. Wansbrough, 'Ugaritic in Chancery Pratice,' in *Cuneiform Archives and Libraries — Papers Read at the 30th Rencontre Assyriologique Internationale, Leiden, 1983* (Publications de l'Institut historique et archéologique néerlandais de Stamboul, LVII), Istanbul 1986, pp. 205–209. Further observations on these matters are set out in my forthcoming study entitled *Chancery Practice and the Problem of* Lingua Franca.

[19]M. Silver, *Economics Structures of the American Near East*, London 1985, esp. pp. 71–144 ad theses of K. Polanyi finally expressed in the posthumous edition of H. Pearson, *The Livelihood of Man*, New York 1981; but cf. already K.R. Veenhof, *Aspects of Old Assyrian Trade and Its Terminology*, Leiden 1972, esp. pp. 345–357; R. Adams, 'Anthropological Perspectives on Ancient Trade,' *Current Anthropology*, XV (1974), pp. 239–258; J. Gledhill and M. Larsen, 'The Polanyi Paradigm and a Dynamic Analysis of Archaic States,' in *Theory and Explanation in Archaeology*, New York 1982, pp. 197–229.

historians. To have achieved such diametrically opposed readings, each must have started from an independently adopted 'system of co-ordinates.' Like most of the random and discrete findings of ancient Near Eastern archaeology, the material from Ugarit is obstinately mute. Its organization demands a self-conscious commitment to a style of historical discourse that equates causality with continuity. But it is precisely the absence of continuity in these data that attracts attention to the stylistic exercise. Prosopography is exiguous, localities are elusive, institutions evanescent, and the actual transactions of daily life a matter of deduction from 'common sense.' That despite these disabilities a coherent account of Ugarit could have been produced attests to an admirable and perennial mimetic talent. For the archeologist Aristotle's 'referent' is supplied; its 'context' = significance has got to be found, and that is the reason for my juxtaposition of two such markedly different specimens of historical inquiry.

And yet, their treatment has not been so very different. It must by now have become clear that my expectations of historical method are seldom fulfilled. I should have supposed that two such contrasting sets of data must generate distinctive modes of analysis. Instead, a mildly interesting convergence of method is discernible: while the artifacts of Ugarit have been translated into a narrative pattern of events, the literary account of the Hijaz has gradually assumed the status of an archaeological site. The element common to both is stratigraphic analysis. Its purpose is identification of something tangible that can in turn be called 'fact.' On a dig, this imagery is naturally persuasive; in a chronicle it is in danger of missing the point. But it does indicate selection of a paradigm that generates not merely the appropriate question but also the type of answer expected. Once uttered that expectation is rarely disappointed. It is after all in the nature of things that it should not be. And that is what one might, perhaps uncharitably, call the 'tyranny of history.'

Now, in recent years a great deal (even, perhaps, too much) has been written about the nature of 'historical understanding,' identified by such tags as 'metahistory,'[20] 'dialectic,'[21] and 'hermeneutics.'[22] But no amount of conceptual theorizing has been able to dispel the apparently deep-seated

[20] H. White, *Metahistory — The Historical Imagination in Nineteenth-Century Europe*, Baltimore 1973, esp. pp. 1–42.
[21] F. Jameson, *Marxism and Form — Twentieth-Century Dialectical Theories of Literature*, Princeton 1971, esp. pp. 306–416.
[22] R. Palmer, *Hermeneutics*, Evanston 1969, esp. pp. 3–71; E. McKnight, *Meaning in Texts — The Historical Shaping of a Narrative Hermeneutics*, Philadelphia 1978, esp. pp. 91–204.

conviction that 'history' is essentially historiography. Whatever acts of collection and collocation might precede the composition, its expression is narrative. I am also inclined to believe that its perception too is narrative: that is to say, follows a 'storyline,' has something like a 'plot,' is linear (exhibits causal nexus) and cumulative (everything counts). It is according to these parameters that one can understand the seductive power of sentence structure. Attempts to escape this force are made from time to time, e.g., in 'structuralism'[23] by dimissing the concept of 'referent'; in 'deconstruction'[24] by denying 'syntactic' continuity in experience. Neither has found, or is likely to find, universal assent. The reason for that lies probably in some vague but enduring conviction that the record has got to be readable. And this will be as much a matter of epistemology as of literature. There is, however, another factor in this process, a kind of safety-valve, as it were, that at the occasional expense of readability makes the record manageable: by reducing the cumulative burden and punctuating severely its linearity —

> there is no exercise of the intellect which is not, in the final analysis, useless. A philosophical doctrine begins as a plausible description of the universe; with the passage of the years it becomes a mere chapter — if not a paragraph or a name — in the history of philosophy. In literature, this eventual caducity is even more notorious.[25]

That statement, from one of the greatest contemporary observers of the human condition, can be differently expressed as 'textbook simplification,' i.e., the summary of evidence in the form of detachable conclusions, or the relegation of earlier argument to condensed footnote references. These techniques, by which enormous effort and vast erudition are reduced to manageable proportion, might be described as perennial features, hence constants of the historical record.[26] They are particularly noticeable in the two works I mentioned a moment ago in the context of Ugarit. No one at all familiar with the sources (!) for Bronze Age history could suppress a gasp

---

[23]See references above, in notes 8 and 22.

[24]M. Foucault, *The Order of Things — An Archaeology of the Human Sciences*, London 1970; cf. H. White, 'Foucault Decoded — Notes from Underground,' *History and Theory*, XII (1973), pp. 23–54.

[25]J.L Borges, 'Pierre Menard, Author of the Quixote,' *Labyrinths*, London 1970, pp. 69–70.

[26]Cf. Kuhn (above, n. 7), pp. 136–143; L. Mink, 'The Autonomy of Historical Understanding,' *History and Theory*, V (1966), pp. 24–47.

of astonishment at the occasional genius but persistent audacity of Polanyi and Silver in their recomposition of those laconic materials.

The 'detachable conclusion' is of course a recurrent feature in histories of science. There, apparently, the context of problem-solving matters less than the solution itself as component of an abstract process more or less independent of its historical circumstances. The average reader's knowledge of Einstein's contributions to a general theory of relativity, for example, are seldom conditioned by any acquaintance with his development as a musician, philosopher or Zionist. Despite some recent, and occasionally polemical, contributions the same may be said about historians of the Near and Middle East.[27] This would matter less for the ancient segment of that history, for which we have only archaeological evidence (and its modern exegetes are well known), than for the medieval period, for which we have only literary evidence.

But with that complaint we (or at least I) have now come full circle. My intention was to ask: 'What is obvious, or self-evident?' The answer, you must by now have guessed, is: 'Nothing, nothing at all.' No record is unambiguous, and each demands an informed approach. In a recent and typical assault on this problem, Moses Finley declared a vested interest in the value of historical documentation over archaeological artifact.[28] With that assertion he must have wished to announce a preference for the authorial presence of the chronicler to the inarticulate existence of a chance discovery. To that I can only say that it may *seem* easier, but is in fact the more difficult alternative. Neither kind of witness can of course be properly interrogated. Nor can the circumstances of either be properly reconstructed. Each utterance requires a special sort of exegesis that ought to take the place of a candid but naive appeal to 'common sense.'[29]

In conclusion I should like to repeat a story that is in this company very familiar, but which nonetheless is so stunningly relevant to the caducity of literary transmission that I could not resist:

> When the Baal Shem had a difficult task before him, he would go to a certain place in the woods, light a fire and meditate in prayer — and

---

[27] E.g., E.W. Said, *Orientalism*, London 1978; R.C. Martin (ed.), *Approaches to Islam in Religious Studies*, Tucson (Arizona) 1985.

[28] M. Finley, *The Use and Abuse of History*, London 1975, esp. pp. 87–101.

[29] Valuable correctives in *Biblical Archaeology Today — Proceedings of the International Congress on Biblical Archaeology*, Jerusalem, April 1984, Jerusalem 1985, esp. F.M. Cross (pp. 9–15), B. Mazar (pp. 16–20), Y. Yadin (pp. 21–27), H. Tadmor (pp. 260–268), and E.E. Urbach (pp. 502–509).

what he had set out to perform was done. When a generation later the
'Maggid' of Meseritz was faced with the same task he would go to the
same place in the woods and say: We can no longer light the fire, but we
can still speak the prayers — and what he wanted done became reality.
Again a generation later Rabbi Moshe Leib of Sassov had to perform this
task. And he too went into the woods and said: We can no longer light a
fire, nor do we know the secret meditations belonging to the prayer, but
we do know the place in the woods to which it all belongs — and that
must be sufficient; and sufficient it was. But when another generation
had passed and Rabbi Israel of Rishin was called upon to perform the
task, he sat down on his golden chair in his castle and said: We cannot
light the fire, we cannot speak the prayers, we do not know the place,
but we can tell the story of how it was done. And the story which he
told had the same effect as the actions of the other three.[30]

Now, could there be more eloquent testimony to the imaginative reconstruction of the past? Every author creates not merely his own precursors,
but the very record of their activity, and I should not like to see historians
exempted from this responsibility.

*Prof. John E. Wansbrough delivered the Albert Einstein Memorial Lecture
in 1986.*

---

[30]S.J. Agnon, in G. Scholem, *Major Trends in Jewish Mysticism*, New York 1961, pp.
349–350.